The six titles in this series are:

Seas and Inland Waters

Tropical Forests

Deserts

Grasslands

Mountains and Forests

Polar Regions

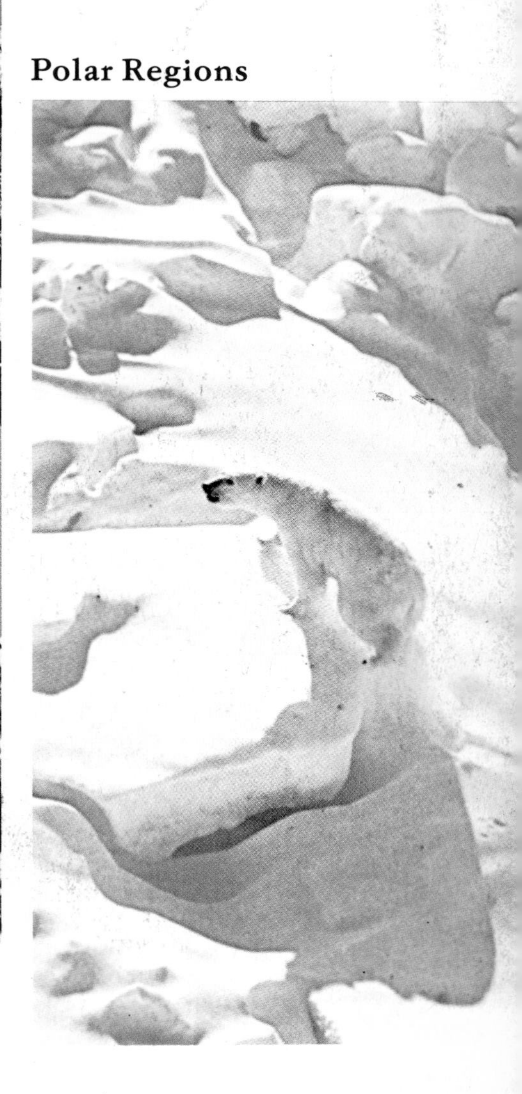

Animals and their Environment

Animals of the Polar Regions

Federica Colombo and Gina Barnabé
Illustrated with drawings by Gabriele Pozzi

Burke Books LONDON * TORONTO * NEW YORK

CIP data
Colombo, Federica
Animals of the polar regions — (Animals and their environment)
1. Zoology — Polar regions. — Juvenile literature
I. Title II. Series
591.909'1 QL104

ISBN 0 222 00853 9

Burke Publishing Company Limited
Pegasus House, 116-120 Golden Lane, London EC1Y OTL, England
Burke Publishing (Canada) Limited
Toronto, Ontario, Canada.
Burke Publishing Company Inc.
540 Barnum Avenue, Bridgeport, Connecticut 06608, U.S.A.
Printed by Vallardi Industrie Grafiche S.p.A. Milan

Acknowledgements

The photographs are reproduced by permission of:

Burton; Bruce Coleman; Dimt; Erice; Explorer; Björn Finstadt/Olav Bjaaland; Fratas; Harris; Hans Huber; Jacana-Boisson; Jonin; David Linton; Lippmann; Maffei; Marka-Afsen/Maltini-Solaini; Milwaukee; Monkmeyer Press Photos; Norway Travel Association; Ott; Lee Rus; Sef; Soper; Suinot; Sundace; Tiofoto; United Press International and Visage.

Contents

The Ends of the World

The Polar Caps

The Arctic or north polar region consists of a permanent ice cap, which covers the Arctic Ocean and is surrounded by the landmasses of the Old and New Worlds. The Antarctic or south polar region is a continent surrounding the South Pole, permanently covered by ice, which in some places is up to 3,500 metres thick. Permanent ice on the Poles is an important factor affecting the climate of the whole world, because the cold currents which cross the oceans all start there. There are some similarities between the Poles such as the extreme cold and the auroras. The auroras are remarkable phenomena caused by the light producing colours and haloes in the sky. These colours vary from red to blue and the effect may last for minutes or for several days with a curtain of colour in the sky. The Northern lights, or Aurora borealis, are a common phenomenon visible all round the Arctic circle and can even be seen as far south as London. The southern aurora is called the Aurora australis.
Another feature common to the two Poles is the alternation of a long day, lasting six months, and a night of the same duration. The major difference between them is that when the sun is shining continuously on the South

In the regions inside the polar circles, the sun stays above the horizon for six consecutive months. The photograph shows the midnight sun at the North Pole. After the long polar day comes six months when the sun never rises.

In the southern hemisphere, towards the end of Spring the edges of the ice plateau break up and drift. In the Antarctic, the violent seas break off enormous icebergs from the Ross Ice Barrier. These are often flat and up to 10 kilometres long. They are carried by currents towards tropical regions, gradually melting away until they finally disappear.

Pole the North Pole is having its six months of darkness. There are, however, many other fundamental differences between the two Poles. To begin with the North Pole is a sea and the South Pole a continent. The South Polar region has ninety per cent of the earth's ice. The Arctic region is flat and open; the Antarctic continent is mountainous. The minimum temperatures at the South Pole are far lower than those at the North Pole and can fall to 90°C. below freezing. This is partly due to the much higher altitude of the Antarctic continent, where the mountains may reach up to 4,000 metres in places. The southern climate is worse: icy, with violent winds which blow continuously, causing tremendous storms. As a result, the Antarctic waters are rough and stormy. In this barren region, plant life is only present on the extreme edges of the Antarctic Peninsula, where the average temperature is just above 0°C.

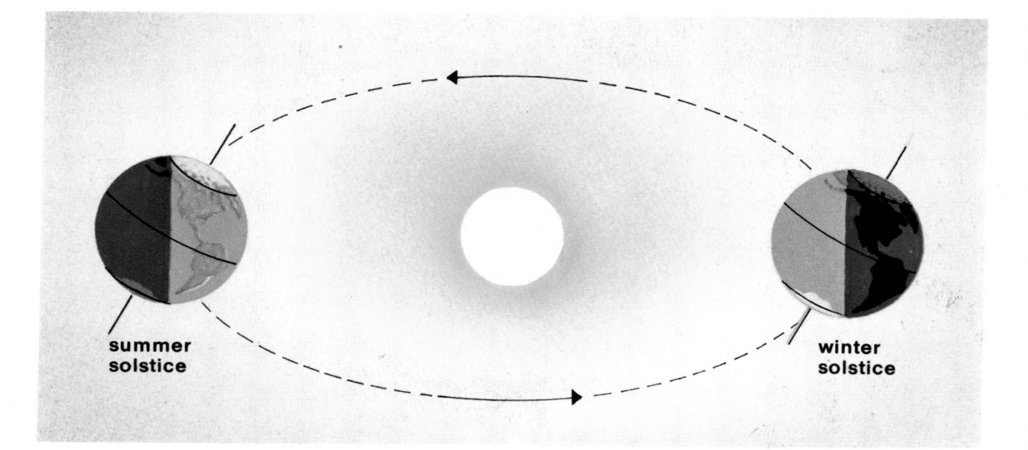

Above: *The variations of the angle of the Earth's axis in relation to ecliptic levels are responsible for the changing seasons. When the sun's rays are perpendicular to the Equator (equinox of 21st March) both Poles are in darkness. During the following three months, the sun moves towards the Tropic of Cancer, where the solstice (21st June) marks the onset of summer in the North and of winter in the South. From this moment the sun begins to descend towards the Equator (equinox of 21st September) and then to the Tropic of Capricorn (solstice of 21st December). That is the time when summer starts in the South and winter starts in the North.*

Below: *The enormous icebergs that plough through the seas of the northern hemisphere come mostly from Greenland. They are broken from the huge mass of continental ice which is about 2,000 metres thick along the edges of the sea. The action of the waves breaks away large towering blocks. With only one seventh of their size projecting above the water, the icebergs are driven southwards by the currents from Labrador and are a constant threat to navigation – so much so that they are subjected to careful international surveillance.*

ARCTIC REGIONS

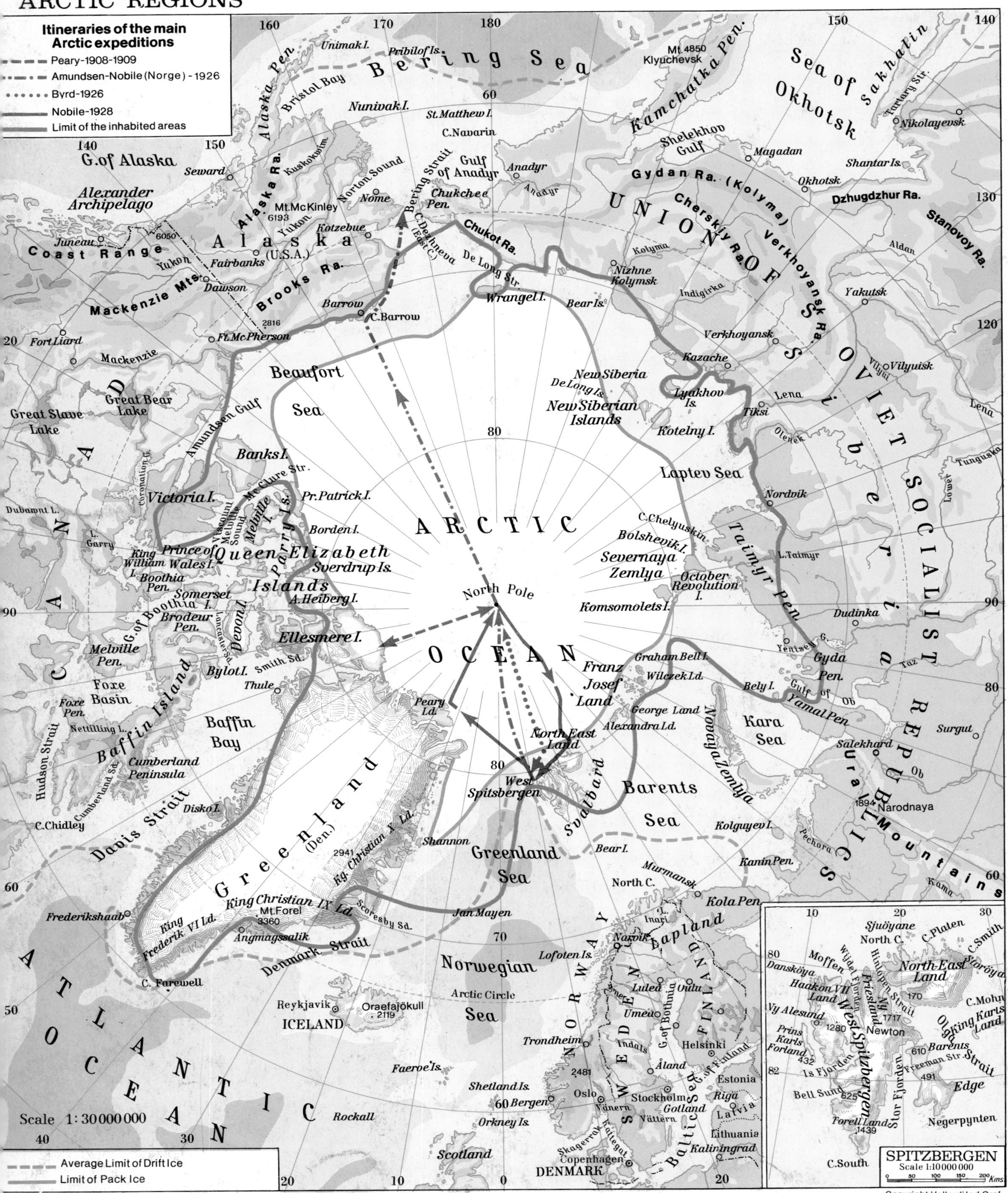

The North Pole

The Conquest of the North Pole

Above: *The American explorer, R.E. Peary – the first man to reach the North Pole.*

Below: *Ships in the Arctic seas in search of the North-West passage. An engraving from 1850.*

The North-East Passage

The search for the hypothetical passages round the north of Europe and America (to speed up trade with the Pacific) first started in the sixteenth century. The English, later followed by Dutch and Russian seamen, were the first to undertake expeditions to find a passage to the North-East. It was as late as 1878 before a Swede named Nordenskjöld succeeded in the enterprise, aboard the ship *Vega*. After leaving north Scandinavia, the ship proceeded towards the East. It stayed iced up in Siberian waters through the winter. The following Spring it recommenced its journey, arriving in Japan after passing through the Bering Straits.

The North-West Passage

Simultaneously, other explorers were looking for a passage to the North-West. On these journeys Baffin Island, the Davis Straits and Hudson Bay were touched. It was during one of these expeditions that John Ross tried for the first time to reach the North Pole. After his unsuccessful attempt many others followed. The British Admiralty offered a prize of £2,000 to the first person to reach the North Pole. The North-West passage was finally discovered by Robert McClure, who crossed the Bering Straits while taking part in a search for the Franklin expedition which disappeared mysteriously in 1845.

The Conquest of the Pole

Having found the sea passages, the expeditions now became predominantly scientific and turned their attention to conquest of the Pole. The North Pole was first reached by the American, Robert Edwin Peary, already famous for several expeditions to north Greenland. He prepared himself for the great enterprise by living with Eskimos for four years, learning their way of life. It was this experience which helped him to survive the hostile environment.

Peary failed the first time although he got very near his goal: 87° latitude North. He tried again in 1908 with the ship *Roosevelt*. He left Cape Columbia on 28th February 1909 and headed north with dog-sledges, reaching the Pole on 6th April in the same year. On his return trip, Peary covered the 900 kilometres in only fifteen days.

New Methods of Exploration

With the complete exploration of the

North-West passage by the Norwegian Roald Amundsen in the years 1903-1906 exploration by sea came to an end. Then the aerial race to the Pole started. In 1914 the Russians tried to find a polar route and their attempts were repeated by Amundsen in 1925, using two seaplanes. Like the first sea and land expeditions, the first flights were difficult and unsuccessful. The first to fly over the Pole were two Americans, Byrd and Bennet. They reached the Pole in a three-engined aeroplane on 9th May 1926.

Within two days, an Italian named Umberto Nobile flew over the Pole in his airship *Norge*, accompanied by Amundsen. Nobile repeated the flight in 1928 in the airship *Italia*, which unfortunately crashed on the return journey. A few survivors – Nobile was among them – were rescued by a Russian cruiser.

The aeroplane quickly proved itself to be the best means for such expeditions and, in 1933, Lindbergh studied the possibility of a commercial air route between California and Moscow via the North Pole. A few years later the first flights were made between Europe and Japan via a polar route.

Finally, in 1958, the American nuclear submarine *Nautilus* accomplished a sensational mission by reaching the North Pole underneath the polar cap. As well as these major undertakings, many other expeditions took place during which teams of scientists were able to make geographical, geological and biological surveys of great interest and importance.

Above: *On 6th April 1909, R.E. Peary reached the North Pole leading a party of only five people: his assistant Matt Henson and four Eskimos. After surveying with a sextant, they held a brief ceremony: Peary erected the American flag and flags of the associations which had helped him in his expedition. Before returning, the party stayed at the North Pole for over thirty hours, making surveys and meteorological observations.*

Below: *The airship* Italia, *commanded by Captain Nobile, completed three exploratory flights in the polar region in 1928. During the last flight (25th May 1928) it was caught in a violent storm and crashed. Only a few crew members were saved, among them the captain. – (from a drawing of the period).*

The Arctic and Sub-Arctic

The Arctic Ocean

Almost completely surrounded by land, the Arctic Ocean covers an area of thirteen million square kilometres. Near the coasts several distinct seas are recognized, among them the Barents Sea, Kara Sea, Laptev Sea and Greenland Sea. The Arctic Ocean is permanently covered with an ice cap. It is only navigable during the height of summer, from the Barents Sea to Cape Celjuskin, thanks to the warming Gulf Stream. Along the American coast the ice covers the ocean for most of the year, making the North-West Passage practically useless. The Arctic Ocean is connected to the Pacific Ocean through the Bering Straits, and to the Atlantic via the Greenland Sea.

Drifting Ice

The ice that covers the Arctic Ocean is between two and four metres thick and its surface is very uneven because of the constant movement caused by winds blowing over the ocean. Such movement, called "ice-flow", is at a speed of about four kilometres per hour. Near the southern border of the ocean, the ice begins to break up into many floating blocks, thus producing "pack ice" – large masses of broken ice which move around in wind and current.

Above: *Colory Glacier in northern Alaska.*

Below: *Icebergs are a great danger to navigation. The terrible tragedy of the* Titanic, *considered unsinkable, is well-known. During the night of 14th April 1912, on her maiden voyage, she hit an enormous iceberg and sank. Of the 2,200 people on board the liner, 1,500 died in the disaster.*

Arctic Lands

Beyond the most northern parts of Europe, Asia and America – the lands included in the Arctic Circle (north of latitude 66° 33') – there are numerous islands. Among them are Greenland, Spitzbergen, Franz Josef Land, Novaya Zemlya, Severnaya Zemlya, and the islands of Wrangel, Victoria, Ellesmere and Baffin. A common feature of these islands is a very low average temperature, with the ground ice-covered for most of the year. There is an absence of trees of any kind and it is only during the hottest summer months when the ice thaws, showing a layer of marshy ground, that moss, lichen and some grasses begin to grow.

The Sub-Arctic

The term sub-Arctic includes all the territories where the average summer temperature does not exceed 10°C and where the average temperature of the coldest months is below freezing-point. Strangely enough, the world's lowest temperatures are not recorded in the Arctic because the seas have a warming influence. It is in Canada and Siberia that some of the lowest temperatures are recorded; in eastern parts they can fall as low as -50°C. Central Alaska, Labrador, Newfoundland and Iceland are inside the sub-Arctic region.

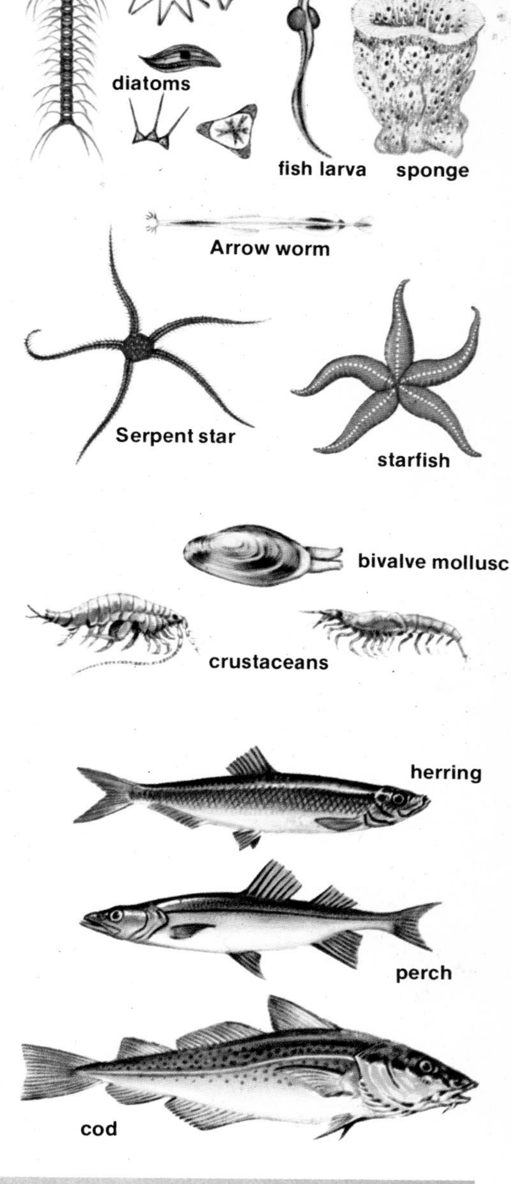

Arctic waters are rich in fish such as salmon, cod and herrings which find abundant food in the zooplankton – the fish larvae and minute crustaceans who, in their turn, feed on the untold millions of phytoplankton. These are single-cell plants, many of which are species of diatoms. The sea also provides food for seals and whales, as well as for enormous colonies of birds which nest on coasts.

Below: *In the Arctic regions, when the snow melts in the Spring, a carpet of bright green moss appears on the soggy ground.*

The Tundra

A Transition Area

The tundra is a vast area without tall trees, stretching between the taiga (a belt of coniferous trees in the northern hemisphere) and the polar regions. The same conditions are found in more southern mountains above the tree-line.

Vegetation of the Tundra

The tundra appears to be a never-ending carpet of grass, moss, lichen and Dwarf Willow (only about 30-40 centimetres high). In the more southerly parts, Birch, Alder, Heather and Bilberry are also found. During the warm season, insects abound. There are three main types of vegetation which can be recognised in the tundra: grassy tundra, wooded tundra and Arctic steppe.

Grassy tundra covers the slopes protected from the wind, where moss and lichen grow, the latter being the favourite food of reindeer, caribou and Musk Ox. In the Spring these areas become covered with brightly coloured flowers: saxifrage, Canterbury Bells, poppies, Arctic Buttercups and campion.

The wooded tundra is found where the terrain is richer in humus because of more frequent rain. Here alder, birch trees, Dwarf Willows and large shrubs form a thick, miniature forest where birds, bears, wolves and large rodents find shelter.

Only a few sparse Dwarf Willows and several types of lichen growing on rocks or gravel flourish in the Arctic steppe.

Top left: *Campion*
Centre left: *Saxifrage*
Left: *Dwarf Willow*
Top right: *Flowering tundra*
Right: *Lichen on a sandstone rock*
Facing page: *Typical Arctic tundra landscape in Greenland*

Formation of the Tundra

Tundra formation can be said to begin with the action of minute algae and bacteria – which isolate nitrogen from the air and transform it into minerals in a layer of windblown debris which covers the permanently frozen ground (permafrost). Once the basi nitrogen-rich soil is produced, then other plants which are resistant to the cold can flourish, and the area becomes richer as time progresses. Later on, earthworms which recycle minerals and organic matter help to produce a more natural soil structure for the establishment of plants. All this is dependent on the establishment of a more fertile soil on top of the permafrost.

The Biological Cycle of the Arctic

The Importance of Plankton

Sunlight is essential to plant life in any area. The green pigment (chlorophyll) of the plant, using the sun's energy, converts inorganic substances into organic ones and provides the basis of life. The stormy nature of the polar oceans and the turbulence of the currents bring rich organic supplies to the surface. With the warmer currents from the south, this mineral-rich sea develops the phytoplankton on which all the animals depend. The richness of the seas in summer, when the presence of birds, seals and whales testifies to the abundance of food, forms a curious contrast to the warmer seas where one might expect an abundance of plankton. This richness of the seas is an indispensible factor in the development of life in the Arctic. Throughout the summer, an abundance of phytoplankton allows the development of zooplankton – microscopic animals which feed on phytoplankton. They in their turn feed the numerous species of crayfish, fish, starfish, birds, seals and whales.

Food-chain

Animals that feed on plankton generally have specially modified mouths provided with organs for filtering water. They are known as filter feeders. Crustaceans and some fish larvae have many hairy filaments in their mouths with which they feed on minute organisms, scooping them up as they swim along. Even some of the large whales have filters. Numerous shoals of fish follow the movement of the plankton on which they feed, thus becoming easy prey for birds, seals and whales. Seals and whales in their turn, are a favourite prey for killer whales, the most ferocious predators of the sea.

On land, the ice-covered wastes are the hunting-ground of the Polar Bear. The bear is the end-point of the food chain, catching seals and fish near the surface of the water. Man, too, hunts in these regions making his own considerable contribution towards the destruction of the fauna, catching seals, whales and many species of fish that live near the surface.

Waste materials – excrement, skeletons and shells – fall down to the sea bed and are attacked by the animals that live there: sponges, bacteria, worms, etc. The remains of these undergo a process of decomposition and putrefaction and this gives rise to organic substances which are used by the phytoplankton. The sun's rays act on these and, through the reaction with their chlorophyll, they give rise to new organic matter which starts the new food-chain.

Plan of the principal biological food-chains in the Arctic and sub-Arctic.

OPEN SEA – PACK ICE

Carnivores of the Tundra

Many carnivorous animals live in the tundra where they find plenty of food amongst the rodents and river fish. Most important are the Brown Bears and in particular the Grizzly Bear which lives in Alaska and Canada. When they stand up they are over two metres tall and weigh 500 kilograms. They are good fishermen, but besides fish they eat berries, buds, honey, insects and eggs, usually not attacking large animals unless they are injured.

Wolves are the most voracious carnivorous animals and travel widely throughout the desolate lands of the tundra looking for food. Considered a dangerous predator, the wolf is ruthlessly hunted throughout its range. Another animal which is always on the lookout for prey and is a threat to smaller animals is the Arctic Fox. It is smaller than the Red Fox and is greyish-brown in colour during the summer but becomes completely white in winter, blending perfectly into the snowy background. This method of camouflage is adopted by many animals that live in areas which are snow-covered for some months of the year.

Small Carnivores

Various small predators, such as otters, mink, marten and ermine have adapted themselves to life in the tundra, finding enough food there and being protected from the extreme cold by their thick fur. In spite of their small size, these animals are ferocious hunters and they skilfully track down the many rodents and other small mammals and birds of the tundra.

Below left: *Perfectly adapted to life in the water, the Sea Otter has an elongated body and webbed feet which enable it to swim with ease. It feeds on fish and crustaceans. Often hunted for its valuable fur, it is at present becoming very rare. It is unusual among animals in that it uses a tool to help with its meals! It floats on its back, with a stone balanced on its upturned stomach. On this stone the Sea Otter cracks open the shellfish which are its principal food.*

Below: *The Polar or Canadian Lynx has a silver-grey coat which is valuable to fur dealers. In appearance it is similar to a large cat and its distinguishing features are the black tufts on the tips of its ears and its thick whiskers. It is very agile, of nocturnal habits and hunts young deer, rodents and birds.*

Above: *Two Grizzly Bears wander in the flowering tundra in search of juicy berries. Also called the Grey Bear, the Grizzly Bear lives in North America, from Alaska to California. In spite of its exceptional strength, it rarely attacks large herbivorous animals and mostly feeds on plants.*

Below left: *Wolves are widespread, although their numbers have been greatly reduced. The remaining ones that live in the north have a light-coloured coat, sometimes almost white, which blends into the background during the long northern winters. Gregarious by nature, wolves hunt in packs.*

Below right: *The coat of the Arctic Fox is completely white in the winter, but in the Spring it changes to greyish brown. The famous Blue Fox is a variety of the Arctic Fox. It has a coat of a delicate slate colour which just becomes lighter in winter.*

Arctic Seals

Above: *This Bearded Seal is lying on a floating slab of ice, basking in the tepid Arctic sun. Between two and three metres in length, the Bearded Seal lives in the Arctic seas and dives deep for food.*

Below: *When the ice begins to form and is still thin, seals make holes in it to enable them to come up for air. Polar Bears and Eskimos may lie in wait near these breathing holes in order to capture the seals.*

Life in the Sea (Pinnipedia)

In prehistoric times, the ancestors of seals lived on land but began progressively to adapt themselves to life in the sea. There are three families of Pinnipedia: walruses, sea-lions and seals.

Like whales, seals have been ruthlessly hunted over the past centuries. Some species were killed for their valuable pelts, others for their reserves of fat. Some were unfortunate enough to be valuable to the hunters for both reasons. All were thought to be in competition with fishermen! Because of this, many species are practically extinct, while others have been saved but only by fairly extensive laws regulating seal-hunting.

Sea-Lions

Unlike seals and walruses, sea-lions have small external ear lobes. The

shape of their fins allows them to waddle along reasonably well on land but they are more graceful in the sea. Some species of sea-lions only live in sub-Arctic seas as their fur is not sufficient protection against the cold, unlike that of seals and walruses. Northern sea-lions form large colonies, often including more than a million individuals. In spite of the indiscriminate killings to which they have always been subjected, they have maintained their numbers. Just before the mating-season starts, males journey north and, on reaching their breeding places, they fight ferocious duels in order to define the boundaries of their territories. After a month the females and young join them. The young sea-lions form separate groups. Soon after their arrival the females give birth and rejoin the group with their new-born. The strongest males which have managed to conquer territories have the right to form harems. One male may have fifty females. During the mating-season the males guard their territories incessantly, not even returning to the sea to eat.
In October, when the young are able to undertake journeys, the sea-lions migrate towards the warmer waters of California and Japan.

Seals

Very agile in the water, seals find it difficult to walk on dry land, where their gait makes one think of a large caterpillar. Thanks to a thick subcutaneous layer of fat that protects them from the cold, seals are able to live in the seas surrounding the Pole, further north than the sea-lions. Typical of the northern hemisphere are, among others, the Grey Seal, the Hooded Seal, and the Harp or Greenland Seal. Seals and sea-lions feed on fish and crustaceans which they catch even in deep waters. In their turn they are hunted by Killer Whales and even Sperm Whales. Finally, the Polar Bear may lie in wait for them when they come up for air through a hole in the ice.

Whales of the North Pole

BEARDED WHALES

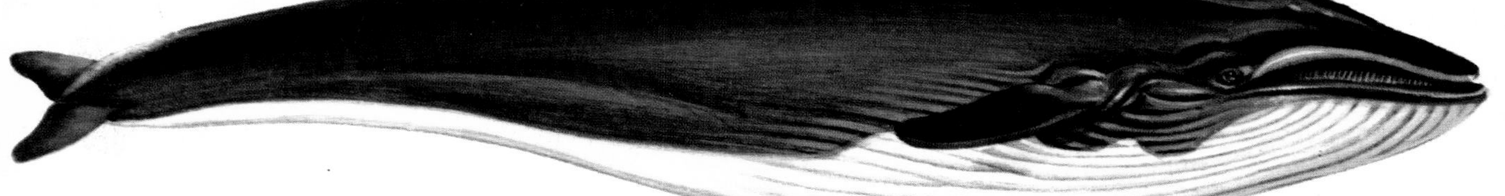

fin-whale

sei-whale

Minke Whale

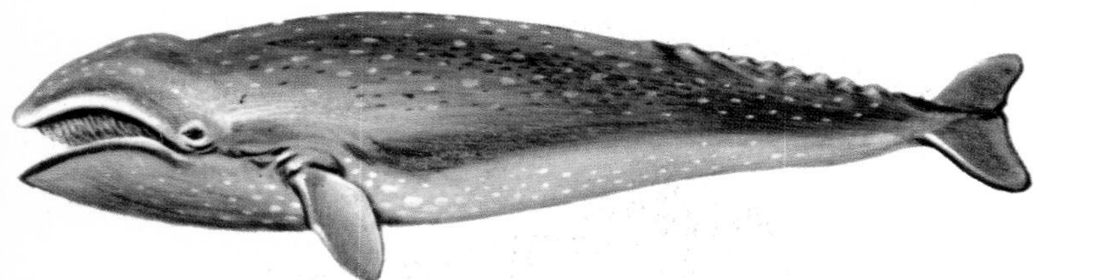

Grey Whale

Humpback Whale

Adapting to Life in the Sea

Whales are mammals which have become completely adapted to marine life: their front legs have been transformed into fins, their back legs have disappeared and the tail has become a strong rudder. Their fur has been lost and their smooth but thick waterproof skin and their nostrils opening on to the forehead all testify to their aquatic life. A thick layer of subcutaneous fat (or blubber) maintains the temperature of these warm-blooded giants of the animal world in the cold waters in which they live.

Whalebone (or Baleen) Whales and Toothed Whales

Whales are divided into two groups: Mysticeti which have baleen or whalebones; and Odontoceti, or toothed whales. Mysticeti is the group containing the larger species, such as the Blue Whale, which is the largest living animal – 30 metres in length. The whalebones or baleen (from which they get their name) form a type of filter through which enormous quantities of water, rich in plankton, are filtered. Much of this plankton is made up of shrimps or krill which are sieved in tonnes by the whales.

Odontoceti (toothed whales) have teeth instead of whalebones. (Some

Blue Whale

Greenland Right Whale

dolphins have up to 260 teeth, certainly the highest number among mammals.) The group Odontoceti, includes the smaller whales, such as dolphins, narwhals, etc. Among these only the Sperm Whale reaches a size comparable to that of the Baleen Whales: 16-20 metres in length and 35 tonnes in weight on average.

An Arctic Unicorn

The Narwhal is unique amongst the Whales in having a long thin spiral tusk, shaped like a lance and sometimes more than two metres long. This is formed from one of the two upper incisors, the other one is lost, as are all the remaining teeth. Narwhals grow up to five metres long and are found often in large schools in the Arctic seas. They feed mainly on squid but will also eat fish.
Another of the northern hemisphere species is the Beluga or White Whale, up to 5.5 metres in length, which lives in the coastal waters of the Arctic seas. The White Whale is hunted for its fat and its skin, the latter being sold commercially as "porpoise hide". Like the Narwhal, it feeds on fish and cuttlefish. Other species of whale, whether Mysticeti or Odontoceti, may be found in both the northern or southern hemispheres.

TOOTHED WHALES

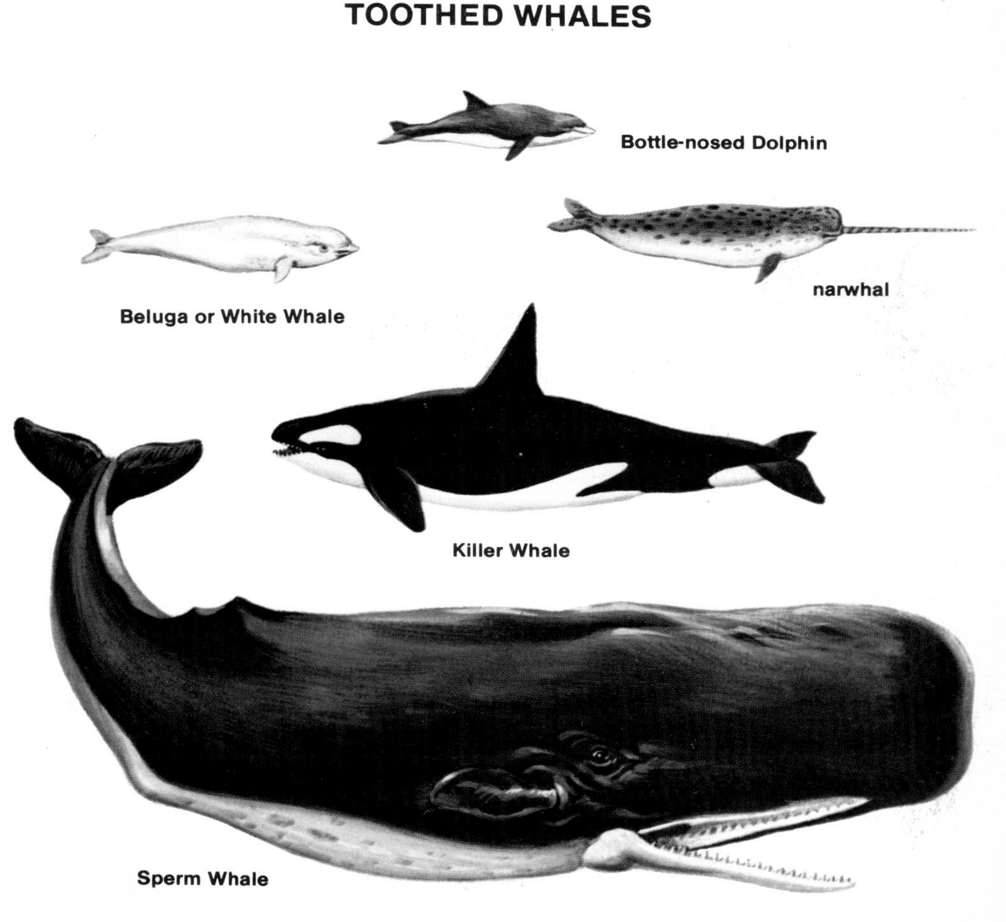

The Herbivores

The Large Herbivores

During the warmer summer months the tundra becomes the goal of some of the herbivores which travel north in search of new pastures. As well as elks, Musk Ox, caribou and reindeer, the tundra is inhabited by several species of sheep, such as the Rocky Mountain Sheep from Alaska and related species in Eastern Siberia, and the mountain sheep of northern Canada. They are well adapted to the surroundings and manage to survive in spite of a scarcity of food in the sub-Arctic regions. During winter they dig in the snow with their hooves in search of grass.

The Small Herbivores

Hares, lemmings, voles, marmots and other small rodents inhabit the vast and desolate northern areas. The numbers of these small rodents undergo periodic fluctuations, which are more or less regular in relation to natural factors, among which is the quantity of food available. When the weather is particularly severe and food is scarce, many die of hunger; but when food is plentiful, they breed rapidly. However, these large numbers of rodents attract many predators which, with an abundance of food, increase in their turn. The result of the increased number of predators is a reduction in the number of rodents. This in itself reduces the amount of food available for the predators and their numbers fall. Subsequently this enables the rodents to increase and the cycle begins again. This rise and fall in numbers of prey and predators in cycles occurs widely in nature.

Above: *These Iceland ponies are particularly hardy and frugal. Agile and intelligent, they are adapted to life in difficult areas. Their long, thick coat helps to protect them from the particularly severe climate of the island.*
Below: *The Arctic Hare which lives in Northern America, has a greyish-brown coat in summer; in winter it becomes completely white. In comparison with the common species, it has smaller ears to prevent excessive heat loss in the cold weather. The same applies to the Arctic Fox and other Arctic species.*

Below left: *The Pika or Mouse Hare, about 50 centimetres long, belongs to a family of Calling or Whistling Hares, so-called because of the constant bleating noise which they produce when communicating with others of the species, or warning of danger. They live in the Himalayas and in the Rocky Mountains of North America.*

Below centre: *Marmots like this one spend a large part of their lives sleeping; their hibernation lasts between six and nine months.*

Above: *The Wapiti is sometimes called an elk in North America, although it is different from the true elk* (Alces Alces). *It is larger than the European deer and is about 1.70 metres high and weighs 500 kilograms. As a result of the indiscriminate hunting to which they have been subjected for centuries, Wapiti are now greatly reduced in numbers and found almost exclusively in national parks.*

Below: *Lemmings are small rodents (about 15 centimetres long) from the sub-family Microtinae which includes voles and muskrats. They live in northern Europe and Arctic Asia and are very prolific. They are the main food of foxes, Snowy Owls and other predators. In the years when they become particularly numerous, Lemmings undertake mass migrations during which many die trying to cross rivers or as victims of predators.*

The Large Bird Colonies

Coasts, Swarming with Life

In the warmer months the coast of the Arctic and Antarctic seas are colonised by thousands of sea-birds. Far from human settlements, the birds nest in peace. The cliffs

and small islands are inhabited by a number of species of sea birds which form large colonies, sometimes numbering several thousand individuals. Gannets, guillemots, razorbills, cormorants, gulls and many other species have specialised in building their nests even on the narrowest ledges and most precipitous cliffs overhanging the sea. Nests, if any, are built very close

The coasts of northern Europe, America and Asia are inhabited by Guillemots – large sea-birds which vaguely resemble penguins. They are about 40 centimetres long with elegant black and white plumage and they live in large colonies on rock ledges. They do not build nests. The eggs are deposited on the rocks and incubated by both parents. The eggs have a narrow end and a broad end so that, if accidently kicked by the parent birds, they will roll around rather than fall off the ledge.

together and every pair defends its own territory from any intrusion. The birds are capable of finding their nests immediately on their return from every flight. The huge bird colonies can be easily seen from ships even from considerable distances. The coasts are bubbling with life. Some birds are arriving, some are flying away in search of food; and some are arguing about the limits of their territories.
All these birds exploit the sea for their food. Larger species are excellent divers and pursue their prey – mostly fish, crustaceans and molluscs – even to depths of 100 metres: smaller ones, however, are satisfied with what they find near the surface.

Above: *Gannets are large sea-birds which may grow to 95 centimetres in length. They nest in vast colonies on small islands near the coast. Nests are built very close together, containing only one egg incubated by both parents. Because of their size and exceptional wingspan, the gannets take off from the ground with difficulty and prefer to let themselves fall from a rock or rise into the wind and thus take flight. When taking off from the sea, they first run on the surface of the water, flapping their wings like geese and ducks. They acquire their full adult plumage in their fifth year. The young are rather dark with white spots.*

Right: *The shag nests along the coastlines of northern Europe and the Atlantic islands, on rough rocks overhanging the sea. An excellent swimmer and diver, it pursues the fish underwater, usually surfacing to swallow them. It is almost exclusively marine, unlike its rather similar relative, the cormorant.*

Arctic Birds

Attracted by an abundance of food in summer and by the absolute solitude, many birds nest in the Arctic and Antarctic areas.

A Paradise for Birds

The Arctic and sub-Arctic are the nesting areas of great numbers of birds which migrate to these desolate places in the summer. Most are species capable of long flights, such as the albatross, petrel and fulmar, which spend their lives at sea, flying over it or resting on it and coming ashore only to breed.
Along the Arctic rivers and estuaries, there are many different species of waders. Among these are plovers, curlews and sandpipers, which feed

Above: *Typical of the species living in the tundra is the ptarmigan. Its summer plumage of brownish-red with white spots becomes completely white in winter, blending perfectly into the snowy background. It is protected from the severe cold by the soft, thick plumage which covers even its feet.*

Below left: *The Snow Goose nests in north eastern Siberia, in the Arctic regions of America and in Greenland. Gregarious by nature, Snow Geese form large flocks which fly in a characteristic V-shaped formation on their migratory flights. Their plumage is completely white, with a black tip to the wings.*

Below right: *The Canada Goose is a large bird, nearly one metre long. It is very strong and accomplishes long migrations, flying at great speeds. Before leaving, it moults and during this moulting period it is unable to fly and so feeds on the ground where its only defence is its ability to run. It has been introduced to many parts of Europe.*

on small invertebrates which they dig out of the mud with their long slender beaks.

Many species of ducks winter on the coasts and lake banks in the tundra among them the Eider duck, the King Eider, the Long-tailed Duck (which often nest in colonies) and the Pink-footed Goose.

The many lakes of the tundra are inhabited by divers, or loons, large solitary birds which are ideally suited to life on water. There are several different species, all excellent divers, including the Great Northern Diver capable of diving up to 60 metres below water in search of food. It is not only the large and medium-sized birds who migrate north to breed. Great numbers of small species including larks, wheatears and buntings, also fly north as the humid summer tundra provides innumerable varieties of insects in sufficient numbers to feed the flocks of insectivorous birds. There are some species of birds which are permanent residents of the tundra and do not migrate; these include the capercaillie, the ptarmigan and the raven.

Above: *The Snowy Owl is a large bird which nests in the Arctic tundra. It has white plumage speckled with brown. From a high position, the Snowy Owl surveys the surroundings and when it sees possible prey – usually a small rodent – it takes off and snatches the prey in one swoop like a falcon.*

Below left: *Widespread in sub-Arctic areas, gulls generally nest on sea coasts or the banks of lakes. They are excellent fliers and swimmers and mainly feed on vegetable or animal debris. They are also capable of catching live prey, such as fish and invertebrates. The photograph shows a pair of kittiwakes at their nest.*

Below: *The Sea Eagle, of which there are several species in Eurasia and North America, lives in the northern parts of these continents. The American species which include the Bald Eagle (the national emblem of the United States) nest in rocky gorges in the Arctic and on trees in the taiga. It is a predator of sea birds and fish and migrates south when the first ice appears.*

The Polar Bear

Ruler of the Ice

Undisputed master of the North Pole, the large Polar Bear travels widely across its domain searching for food. It is an enormous white animal which, in spite of its size, is agile on the ground, an excellent swimmer and a tireless traveller. This is a short description of the Polar Bear, master of the Arctic. More carnivorous than other bears, it is the largest and strongest predator of the North and does not hesitate to attack even man. An adult male, on average, weighs about 450 kilograms and is about 2.50 metres in length. Some old males weigh up to 700 kilograms and are 3.30 metres in length.

The Polar Bear rarely reaches land, preferring to live on the ice and often being carried along on ice floes or swimming across wide stretches of sea. It has no fixed lair and only seeks shelter in a ravine or gorge to sleep on the very coldest days. It leads a very solitary existence; only during the mating season do the male and female come together and even then only for a short period. Ferocious battles between two males can take place during this time.

Birth in the Depth of Winter

When the polar winter begins the pregnant females seek a suitable refuge. Usually they dig a hole in the snow in which they sleep. In the middle of winter the cubs are born in this lair, protected from the appalling conditions outside. The cubs, which weigh about 700 grams at birth, are both naked and blind when they are born. They survive in the warmth of their mother's thick fur and feed at her teats for many months. When the time comes to leave their lair the cubs

A mother and her two cubs. Definitely more carnivorous than other bears, the Polar Bear is the largest predator of the far north. Its average length is 2.5 metres and it weighs about 450 kilograms.

have already grown considerably and are covered with thick white fur. They remain with their mother for two years, during which time they follow her everywhere learning the hunting skills which help them to survive in the frozen wastes of the extreme north.

Above: *A thick coat and a layer of subcutaneous fat enables the Polar Bear to survive on the ice and even to swim great distances in the icy sea. The Polar Bear is a tireless traveller, wandering continually on the ice, often allowing himself to be carried along by floating ice packs in search of Seals – his favourite food.*

Below: *Already a rare species, the Polar Bear is now protected by strict conservation laws and can only be hunted during certain times of the year. Naturalists carry out complex studies of its behaviour patterns and distribution. In order to ascertain their numbers and their state of health, scientists drug the bears using guns which fire anaesthetic darts.*

An Expert Hunter

Protected by its coat which blends into the snow, showing only its eyes and black nose, the Polar Bear gets right up close to its prey by crawling on its stomach rather like a huge cat, and then suddenly springing at its victim. Of all its prey, the Polar Bear prefers the seal, particularly the Ringed or Jar Seal. In winter, when the sea is completely frozen over, the Polar Bear waits patiently by the breathing holes for a seal to appear. With a strong blow it breaks the seal's back and drags it from the hole. When food is scarce, small mammals, eggs, algae and even carrion have to suffice.

The Walrus

Above: *It is unusual in carnivorous animals for the teeth or tusks to continue growing; those of the walrus are the exception.*

Below: *During their long migrations, walruses often rest on ice floes.*

Facing page: *An enormous herd of walruses basking in the Arctic sunshine.*

Found only in the Arctic Ocean, the Walrus is by far the largest seal of the north polar region: it is four metres in length and weighs 1.5 tonnes.

In addition to the tusks, common to both males and females, the walrus is easily identified by its head which is disproportionally small in relation to its enormous body, and by its wrinkled skin which is several centimetres thick. Beneath this skin lies a thick layer of fat which provides an excellent protection against the rigours of the Arctic climate. The walrus is believed to be rather limited in intelligence.

Of all its senses only that of smell is really well-developed and this is used to anticipate and give warning of the approach of an enemy. The walrus lives exclusively near coasts and finds food in shallow water. In winter, it migrates south but stays within the limits of floating ice, returning in summer to the north. The walrus is a very good swimmer and can achieve speeds of over 24 kilometres per hour.

Born in the Sea

Walruses breed only once in every two years. After eleven months gestation the female gives birth to her young in the sea. The new-born baby is 120 centimetres long and weighs about 50 kilograms. As soon as it is born the baby attaches itself to its mother's neck. She feeds it for two years and defends it courageously against all dangers. Killer Whales and Polar Bears often attack the young walruses, not having enough courage to confront the adults. Young walruses are very sociable animals and live in large groups, but when they reach sexual maturity they tend to form family groups consisting of one male and two or three females and their young.

Multipurpose Tusks

The impressive tusks, straight in males, smaller and slightly curved in females are very valuable tools for walruses. The tusks grow throughout the animal's life and some old walruses have tusks one metre long. With these overgrown canines, walruses plough the sea bed in search of molluscs such as clams and other bivalves, sea-worms and fish on which they feed. They locate them with their large silken whiskers which provide a sense of touch. The tusks help the walrus when it is climbing and walking on ice slabs and they are also used in defence.

The Alaskan Fur Seal

Compared with other seals, the Alaskan Fur Seal is rather small and graceful. Its fur is thick, shiny and silky and is most valuable.

Valuable Fur

The Alaskan Fur Seal is rather small compared with other seals: males are 2.30 metres long and weigh about 300 kilograms. This seal has soft, thick silken fur which is very much sought after and even in the past these seals were killed in great numbers. Then, at the end of the nineteenth century their breeding-sites were discovered and, as a result, they were killed in such large numbers that they were brought to near extinction. Only the energetic intervention of countries whose coasts border the breeding areas (the United States of America, Canada, the Soviet Union and Japan) saved the species. Now, in fact, it is forbidden to disembark or even sail close to the Pribilof and Commander Islands without a government authorisation and hunting can only take place during a specified short period in the year.

Migrations

During migration, Alaskan Fur Seals travel as far as California, moving in small groups and never going very far from the coastline. Some, however, travel across the Pacific Ocean to Japan and the North Pacific.
At the start of the breeding season, at about the end of May, the males start the return journey to the Pribilof and Commander Islands — very small places to find in the vast seas of the northern hemisphere. A month later they are joined by the females.

The Formation of Harems

As soon as the first Alaskan Fur Seals land, the islands begin to resound with loud shrieks. Their first impulse is to contest territories with fierce duels. When the limits are finally determined, the males wait for the females, scarcely taking their eyes off their rivals: always ready to push them back if they cross the boundary. A month later the females arrive and the harems are formed, each containing about fifty individuals. Two or three days after their arrival, the females give birth to their young (conceived a year before). They feed them for three months and then leave them to fend for themselves.
Weaning brings many difficulties for young seals: some may die of intestinal infections, many become victims of Killer Whales; and as the pups have the most valuable fur they are at great risk from the hunter.

A Very Varied Diet

Studies completed by naturalists have revealed that Alaskan Fur Seals have a very varied diet which includes herrings, salmon and mackerel, thus bringing them into conflict with fishermen who obviously do not show them much sympathy.
Pebbles of various sizes have been found in the stomachs of seals. This has caused a great deal of speculation over the years. There are many theories put forward to account for it. One idea was that the stones were swallowed to help reduce the irritation caused by intestinal worms,or that they were helpful in the digestion of food swallowed whole. Another theory was that they added weight and made diving easier!
One of the most recent suggestions is that seals swallow stones during long periods of starvation in order to combat the pangs of hunger. It seems, however, that we do not really know why this happens. As with so many animals, a detailed study of their biology is needed.

Above: *Once at the breeding-sites, harems are formed composed of one male and several females. The male jealously guards its own territory and discourages any intrusion by rivals, with continual shrieking.*

Right: *The flippers of Alaskan Fur Seals are covered with a very fine soft skin.*

The Elk

Above: *The giant of the deer family, the elk can grow up to two metres in height and three metres in length, excluding the tail. It is, on the whole, a rather awkward-looking animal. The characteristic palmated antlers are absent in females and immature males (up to two years old). Its coat is generally dark brown on the body and dark grey on the legs.*

Below: *The elk lives in areas surrounding the Pole. There are many species, all rather similar. During the winter they make their home in the conifer forests. In the warmer summer months they migrate to the southern borders of the tundra where they find plenty of food in the numerous marshes. They immerse themselves in the water to escape the biting insects, as do large herbivores in other regions of the world.*

Moose or Elk

The moose or elk can measure two metres at shoulder height and weigh up to 500 kilograms. There are many species, all quite similar, living in Scandinavia, Russia, Canada and Alaska. The male elk may be distinguished from other deer (apart from its size and rather ungraceful appearance) by its enormous, generally palmated, antlers. Like other deer, the elk sheds its antlers in December. They regrow in the Spring, covered with a soft skin called "velvet". When the antlers finish growing the "velvet" comes off in shreds. In order to free itself from the shreds, the elk rubs the antlers against trees or on the ground, a habit common to many deer species. In the summer months, the elk leaves the conifer forests where it found refuge during the winter. It migrates to the southern borders of the tundra — by then full of lush vegetation — and to the marshes where it can find its favourite food.

An Irascible Character

The elk is an animal with an irascible and pugnacious nature, especially in the mating season. At this time it can even be a danger to Man. In spite of its size, strength and vigour, the elk has many enemies, among them the lynx, bears and wolves. From these it may manage to escape by using speed or, if outpaced, its fearsome antlers. Its worst enemy is Man, who has hunted the elk for centuries often using dogs specially trained for the hunt.

Facing page: *The elk's antlers begin to grow in the Spring and are shed in the winter, like those of other deer. An old male's antlers may weigh up to 20 kilograms each with a span of two metres. The size and number of branches can be a guide to the age of the animal.*

The Caribou or Reindeer

The caribou is a typical inhabitant of Greenland and the Canadian tundra. This animal is often considered a sub-species of the European reindeer, which nowadays lives only in semi-domestic conditions. The caribou is large — up to 1.5 metres in height and about 300 kilograms in weight — and is very well adapted to its surroundings. It has a thick coat which acts as insulation against northern winters. Its wide, splayed hooves give it good support on snow and in waterlogged ground. The caribou and reindeer are the only deer in which both sexes have antlers; in this case they are very long and have many points.

The caribou, the American sub-species of the European reindeer (although this distinction is disputed) lives in a wild state in the vast open spaces of the tundra and makes two migrations a year – in Spring and autumn. Experiments are in progress to discover whether it is possible to domesticate the caribou, using the knowledge and experience gained with the European species which have been semi-domesticated for centuries.

Impressive Migrations

Canadian caribou accomplish great migrations of more than a thousand kilometres to reach their winter quarters, returning to the tundra in the Spring. Throughout the summer they live in groups of about a hundred individuals. Then, as the first frosts signal the onset of winter the various groups unite to form enormous herds which begin the long journey towards the southern forests. During migration the caribou always follow the same route, marked by their ancestors centuries ago.

The mating takes place just before the start of the migration. The young

are born in the tundra where there is plenty of fresh food available shortly after the return in the following Spring.

Above: *When the migration time arrives, different groups of caribou unite into impressive herds. During the long journey (which may well be more than a thousand kilometres), the caribou overcome many obstacles, swimming with ease across the many rivers which bar the route.*

Below: *The caribou is larger and more majestic than the reindeer which seems squat, perhaps because of the many years of domestication. The antlers, which are present in both sexes, are larger in the caribou than the reindeer. This also helps to make the caribou more imposing than its European relative.*

The Hunting of Caribou

During the nineteenth century, caribou numbered many million and this annual migration was of great benefit to both North American Indians and Eskimos who obtained food and skins (for clothing and tents) from these animals. Wolves and foxes also followed the large herds; the wolves to attack and kill the weaker members of the Caribou herd; the foxes to feed on the left-overs.

Modern firearms used by Indians and Eskimos, as well as by many white hunters, had a considerable effect in reducing the numbers of caribou. The present numbers are estimated at a few hundred thousand. This reduction has impoverished the native population of Indians and Esmimos. It has also had a marked effect on the numbers of the carnivorous animals who depended on the caribou.

The Musk Ox

The Musk Ox manages to survive in a dry and apparently barren habitat which is frozen for most of the year. When the tundra is frozen this large animal, which is related more closely to the goat than to the ox, scrapes the ground with its hooves hoping to find some lichen on the bare rocks beneath.

Found in northern Canada and Greenland, the Musk Ox is sometimes called a living fossil. These animals are of similar structure to the European and Asian Musk Oxen which became extinct in prehistoric times.

Although the Musk Ox belongs to the sub-family which includes goats, its large size (2 –2.6 metres long and 1.5 metres in height), and the shape of its body and horns make it look more like an ox. It has long, compact fur (thicker than that of any other animal) which protects it from the cold as it wanders continually in search of food.

In spite of its large but rather squat appearance, the Musk Ox is agile, climbing steep slopes with good balance and capable of running at great speed. When threatened they make a barrier, in the form of a semi-circle, with their young in the middle. Although this may successfully protect the young against wolves and coyotes, the closing of ranks is fatal against men with modern firearms who can then easily kill as many animals as they like. The Musk Ox is very trusting by nature. Men can approach solitary individuals without causing any alarm. Musk Oxen still have not learned to fear men although they have been hunted for centuries. When the tundra begins to freeze many Musk Oxen undertake migrations south in search of new grazing but others (mostly those which live in the more southern areas) never leave their region, making do with the sparse food they can find by scraping the snow off the ground with their hooves.

The People of the North Pole

The reindeer is still the principal economic resource of the Lapps, the nomadic herdsmen that live on the Kola Peninsula and in the northern regions of Scandinavia. Besides providing meat, milk and skins, the reindeer is used for carrying loads and for haulage: a pair of reindeer pulling a sledge is capable of travelling 80 kilometres a day with a load of 200 kilograms.

The Lapps — Nomadic Herdsmen

The Kola Peninsula and the northern parts of Finland, Sweden and Norway are inhabited by a race of nomadic herdsmen — the Lapps. Their origins are controversial: some say they should be considered Samoyeds; others that they are Finnish; and some that they are the last surviving members of the primitive inhabitants of Scandinavia.

Reindeer are the primary resources of the Lapps, who use them as a source of food and clothing and for travel and as a form of currency for barter. In summer, whole families of Lapps follow the herds in order to obtain food for the autumn and winter. They are experts on reindeer behaviour and use the animals most efficiently. When a reindeer is killed, the meat is dried, and the skin is tanned and used partly for clothes and housing and partly for sale.

The Eskimos

The Eskimos are an ancient race, perhaps coming originally from Siberia. They live in the Arctic regions of Alaska and Greenland. Mainly nomadic, they travel widely in their search for food and less severe climatic conditions, using dog-sledges and kayaks. The Greenland Eskimos spend the winter months in igloos — houses made from shaped blocks of snow. At other times they live in tents made out of walrus skins and other hides.

Eskimos are divided into clans and patriarchal family groups, although women have an important position in the group. They do have individual possessions but they divide equally between the groups the proceeds of any hunt.

In spite of their precarious and difficult existence, Eskimos are friendly, generous, hospitable people, usually smiling in spite of living in what must be the worst climatic conditions in the world. In the twentieth century, the Eskimos who live in the more southern areas have come into contact with Western civilization and have adopted some of its customs.

The traditional costume of the Lapps is brightly coloured, mainly in blue and red. The lives of these people are closely dependent on the reindeer, whose migrations in search of fresh grazing they constantly follow.

Above: *The kayak, a light boat about five metres long, is made of sealskin stretched over a frame of whalebone. In the top of the kayak there is a round hole for the fisherman to sit in. He tightens the sealskin around himself with a stout cord. If the kayak overturns it can be brought upright again with a single stroke of the paddle.*

Below: *In winter the reindeer herds are followed only by a small specialized group of Lapps. In summer whole families are seen camping near the grazing-grounds. The tents of the Lapps are made of reindeer hide and slightly resemble the tepees of American Indians.*

Below: *An Eskimo woman collecting some of the scarce vegetables which make up part of her family's diet. Hunting and fishing provide most of their food. Eskimos lead nomadic lives, protecting themselves from the weather in igloos during the winter and in the summer in tents made from reindeer hide.*

Hunting and Fishing in the North

An Eskimo family wearing traditional costume.

Patient Hunters

Hunting and fishing are the main activities and source of income of the Eskimos. They use their traditional methods of hunting, either sitting in a tied kayak and fishing, or on land using dog-sledges. A good team of dogs can pull a sledge for eighteen consecutive hours at speeds of about 30 kilometres per hour. Dogs are indispensible to the Eskimo hunters. Their favourite catches are seal and walrus which they hunt either on the ice or in the sea. From these animals they obtain meat, heating-oil and clothing. White men have also exploited the Arctic seas. During the nineteenth century, seals were hunted indiscriminately and many species which had been numerous were threatened with extinction. Only strict international protection laws managed to end such massacres. Canadian Eskimos also hunt Polar Bears, a very coveted catch as it provides a great quantity of food and fat, and a thick warm coat. In addition, the Eskimos hunt caribou (with traps) and birds. To catch the birds they use an implement made of several balls tied with a long cord. This is spun round and round and thrown at the birds rather like the bolas used by South Americans.

Protecting the Rich Sea

Eskimos fish with harpoons in shallow water through holes cut in the ice, or by taking their kayaks out into the river estuaries. The fish is cleaned and dried in the open air to preserve it. The low temperature of the Arctic has one advantage; it helps to preserve meat and fish in a safe condition.

Naturally these areas which are rich in fur-skinned animals such as seals and foxes, and the waters full of salmon and salmon trout, have attracted other men, who have exploited these lands, decimating many species of the native animals.

Along the coasts of Greenland and Norway there are many fish canning factories which are supplied by fleets of large fishing-boats. There are even factory ships which work preparing and freezing fish while out with the fleet.

This large-scale and industrialized fishing could develop into a systematic destruction of the wealth of the seas, whose abundant riches could soon disappear if Man is too greedy.

Above and below: *When the sea begins to freeze, Eskimos cut holes in the ice while it is still soft to enable them to fish even during the long Arctic winter. The holes are visited often to make sure that they do not freeze over.*

Sealskins are dried in the open air, stretched over special wooden frames. These skins have many uses. After being specially treated they are used for making clothes, kayaks, tents, etc.

Arctic seas are rich in valuable food such as salmon and salmon trout. The fish are cleaned and hung out to dry in the open air. Arctic waters are greatly exploited by the large fishing fleets which supply the canning and fish-freezing industries.

ANTARCTIC REGIONS

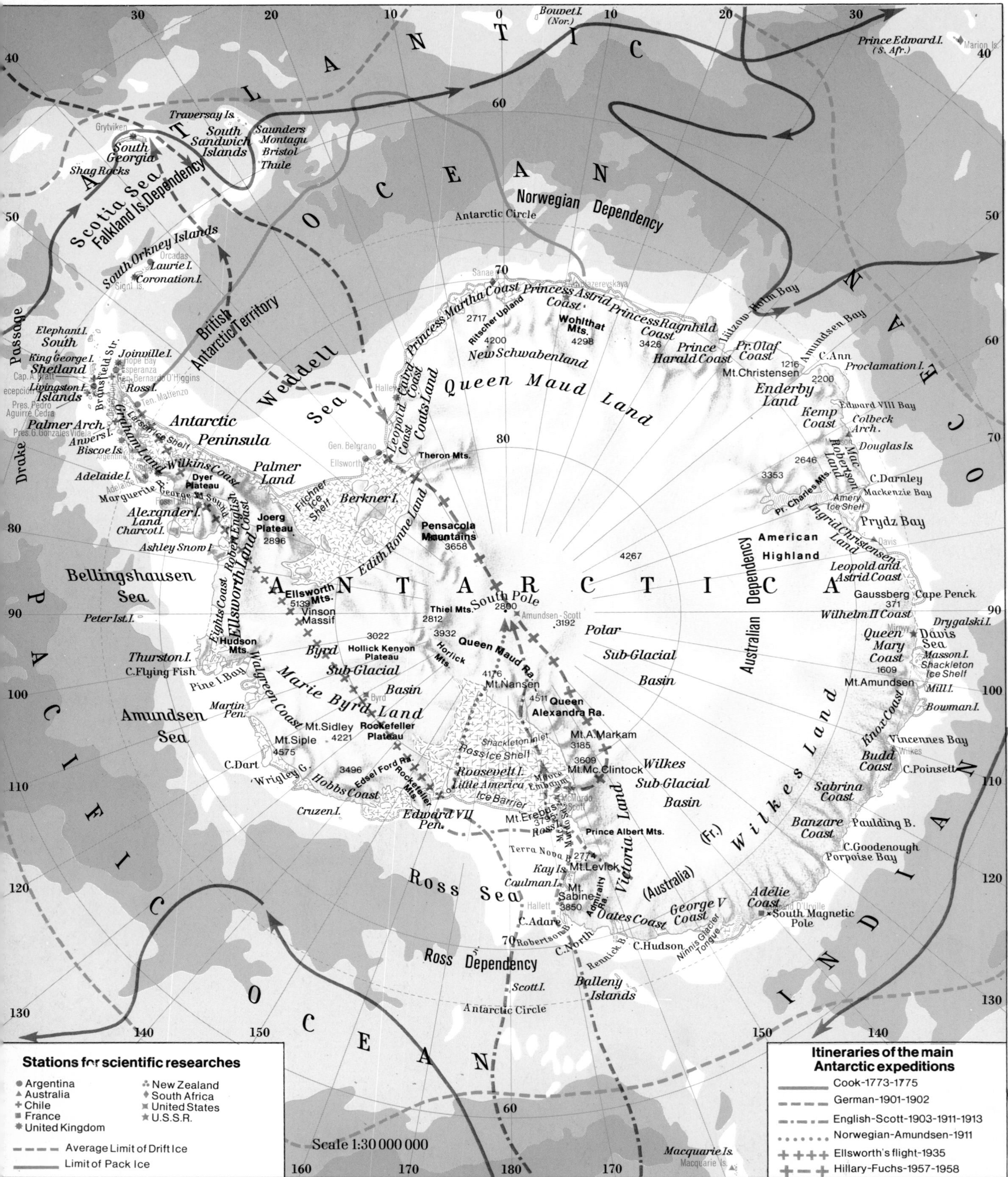

The South Pole

The Conquest of the South Pole

An area outlined by the national flags of the countries which signed the agreement for the peaceful use of Antarctica, indicates the geographical position of the South Pole.

It was Captain James Cook, the English eighteenth-century navigator and explorer, who set out to find land in the South Pacific. It was believed by geographers that such land existed but it had never been seen by western man. On his second voyage, in 1772, Cook left England with two ships, the *Resolution* and the *Adventure*, and sailed into the Antarctic Ocean via the Cape of Good Hope. He went as far as the edge of the icefields, finally sailing to New Zealand before returning to England without actually sighting the mainland of Antarctica.
Between 1818 and 1840 there were many expeditions and many islands and other lands (including Alexander Land and Adelie Land) were discovered. Between 1839 and 1843, during several successive expeditions, James Clarke Ross discovered the McMurdo Sound and the Great Ice Barrier (the Ross Ice Shelf).
At the beginning of this century, Captain Robert Falcon Scott and Sir Ernest Shackleton completed expeditions across the Ross Ice Shelf. Following this, in 1908-1909, Shackleton pressed on towards the interior of the continent to 88° 23′ latitude South, but a series of misfortunes prevented him from actually reaching the Pole.
In 1910-1911, there were two expeditions arranged almost at the same time, specifically for the conquest of the Pole: one was led by the Nor-

Captain R.F. Scott. He explored the Ross Ice Shelf, but later died (with his men) on the return journey from the South Pole, which he reached a month after Amundsen's successful expedition.
Left: *Scott's hut at his base camp at McMurdo was found in 1955. After sixty years it was perfectly preserved.*

wegian explorer, Roald Amundsen and the other by Captain Scott. On 14th December 1911, Amundsen reached the South Pole, after an exhausting march from his base camp in the Bay of Whales, using dog-sledges as his means of transport.

In the meantime, Scott was travelling from the McMurdo Sound towards the Pole. He was using two motor-sledges and Manchurian ponies as he had no faith in the dogs. Unfortunately the mechanical transport soon broke down and the ponies could not endure the cold. Scott had to go back for the dogs, losing precious time. When he reached the halfway point, he realised that he had been beaten by Amundsen. The return journey was even more disastrous and Captain Scott and his men, exhausted by their efforts in the cruel climate, all died. Subsequently, the use of aeroplanes reduced the need for land expeditions and in 1935 many flights of discovery were made over the entire continent from Graham Land to the Bay of Whales.

Amundsen's tent at the South Pole

Right: *Roald Amundsen, born in Borge, Norway, in 1872, died in the Arctic Ocean in 1928. Abandoning his medical studies at the age of twenty-one he dedicated his life to voyages of exploration. He took part in many expeditions including one through the North-West Passage (discovered by Robert McClure in 1850). On this expedition, which ended in 1906, Amundsen spent three consecutive winters frozen in by the ice. When he learned about the conquest of the North Pole by R.A. Peary, he abandoned plans for a similar expedition and turned his attentions to the South Pole, where Scott was already headed. Amundsen reached the Ross Ice Shelf on board the* Fram *in January 1911. He left his base in October and reached the South Pole on 14th December 1911.*

A Frozen Continent

Above: *Marie Byrd Land. The Antarctic is ice-covered. Only parts of the coast furthest from the South Pole are freed from ice in the warmer months. Here moss, lichen and some grass will grow before the onset of winter.*

Below: *Corbett's Bay. Even in summer there are ice-floes on the melting sea.*

The Antarctic Ocean

In 1845 a commission led by the London Geographical Society gave the name "Antarctic Ocean" to the waters south of the Atlantic, Pacific and Indian Oceans, inside the polar circle of 66° 33′ latitude South. When the existence of a continent was established with precision, the name was changed to Australian Ocean and its limits were fixed in 1923 by the Hydrographic Bureau of Monaco as being inside the 55th parallel, but this was not generally accepted. Strong easterly winds are one of the characteristics of the Antarctic producing strong sea currents such as the cold Antarctic currents which reach the south-eastern coasts of Africa, Australia and America, having a considerable effect on the climate.

In its icy waters the current carries away large masses of rich, sub-Antarctic water, rich in living organisms, providing good fishing in the areas where it flows. The Antarctic seas are rough because many minor currents are caused by the movement of surface water towards the north, and of deep waters from the north to the south. The seas surrounding the continent are covered with pack-ice — formations of huge ice-blocks, their sizes varying with the seasons.

An Almost Sterile Continent

The Antarctic, where ninety per cent of global ice is concentrated, is almost completely covered with ice. However, there are areas that are free of ice — these are the lower parts (especially those which extend further northwards away from the South Pole) and the highest mountain peaks (which are actually free from ice). This land, sterile and rocky, is battered by icy winds and blizzards — conditions which greatly limit the growth of plant life. Only some lichen grows in the sheltered parts where the climate is slightly milder. The almost total absence of vegetation means that no herbivorous animals can be supported and therefore there are no land predators. The Antarctic remains the uncontested kingdom of birds, the most characteristic of these being penguins.

Biological Cycle in a Microcosm

The small areas of moss and lichen — usually no larger than two or three metres across — are inhabited by a multitude of insects and bacteria which are part of the food-chain of the area; the herbivores are insects, not large animals. These in turn provide food for birds and predatory

insects. The insects feed on the fragments of vegetation available to them during the summer months. The cycle is similar to that of the Arctic but the animals involved are minute.

Above: *The Antarctic is very mountainous. The highest peak is Mount Vinson (5,139 metres). From these mountains enormous glaciers move slowly towards the sea.*

Below: *The Antarctic is the unchallanged kingdom of sea birds. Many species of penguin, skua, albatross and petrel nest on the coasts, and some even nest hundreds of kilometres inland on the mountain peaks.*

The Great Ice Barrier

The Antarctic continent has two deep inlets, the Weddell Sea and the Ross Sea, parts of which are always frozen over. These enormous masses of ice are called Barriers — the Ross Barrier and the Weddell Barrier. The Ross Barrier, 900 kilometres across and 800 kilometres long, is called the Great Barrier and it forms a triangle, with its point towards the South Pole. Powerful waves break off huge icebergs, sometimes ten kilometres in length, which are then driven northwards by currents until they melt away in the warm tropical waters.

algae

protozoa

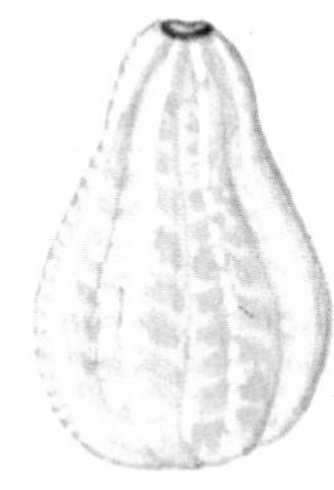

The Sea— A Source of Life

crustaceans

cephalopod molluscs

Fertile Seas and Barren Lands

In complete contrast to the Antarctic landmass, the sea surrounding it contains a wealth of life. There are many reasons for this — temperature, high mineral content, and the length of the polar summer. They all help the process of photosynthesis in which chlorophyll (the green colouring substance in plant cells) turns water, carbon dioxide and sunlight into sugars. As a result, the Antarctic Ocean contains a larger number of creatures than any other sea although the variety of species is more limited.

Phytoplankton

Phytoplankton is a collective name for the minute, single-cell plants that

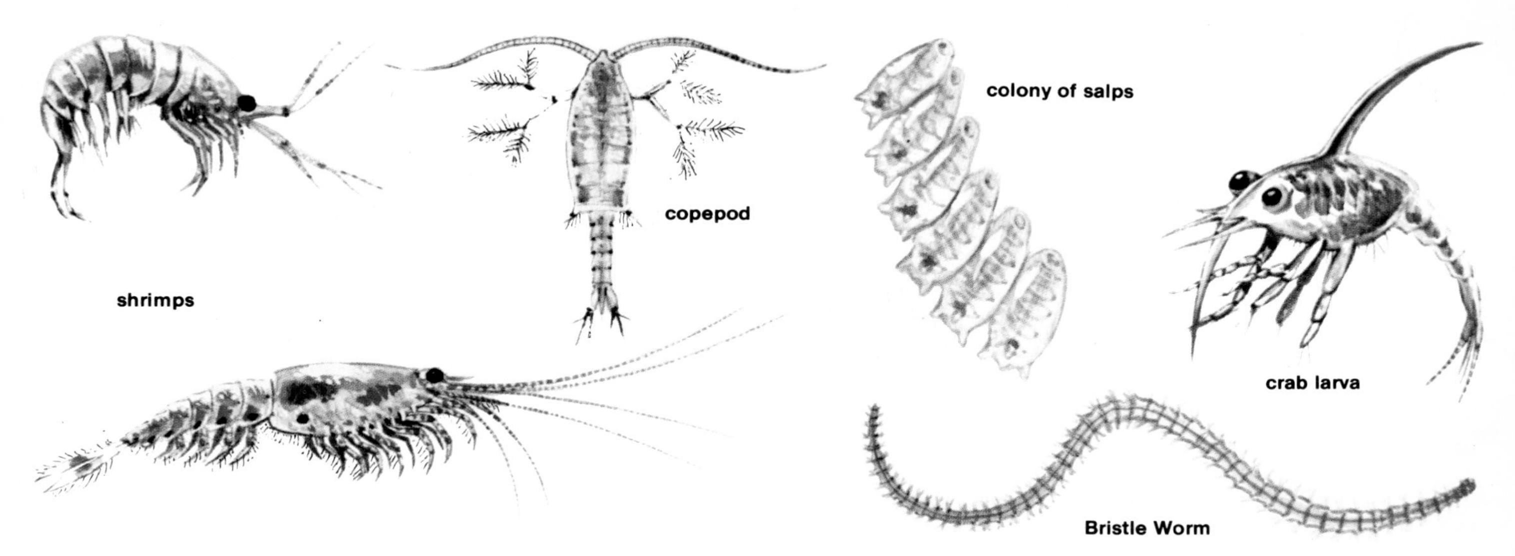

live in the sea. In the Antarctic Ocean the phytoplankton is made up of various algae, each patch composed of millions of individual microscopic plants, living on the surface and colouring the water green and the edges of the ice reddish-brown. Phytoplankton is particularly abundant between the Antarctic Peninsula and the South Shetland Islands.

During the winter, the quantity of phytoplankton diminishes and it sinks deeper into the water. In the Spring, as the ice thaws, the phytoplankton rises to the surface again and grows and multiplies in the sun.

Zooplankton

Zooplankton consists of tiny invertebrates and fish larvae whose life-cycle is closely related to Phytoplankton, with similar annual fluctuations in numbers, decreasing and sinking deep down in winter and rising to the surface and multiplying in the Spring.

Fish and Invertebrates

Many invertebrates enrich the fauna of the Antarctic Ocean: sponges, sea-urchins, star-fish, ascidians, sea-cucumbers, sea anemones and coelenterates. They are joined by many species of crustaceans and molluscs. In spite of the harsh climatic conditions in which they work, scientists have succeeded in identifying sixty species of shallow-water fish and ninety species of deep-water fish.

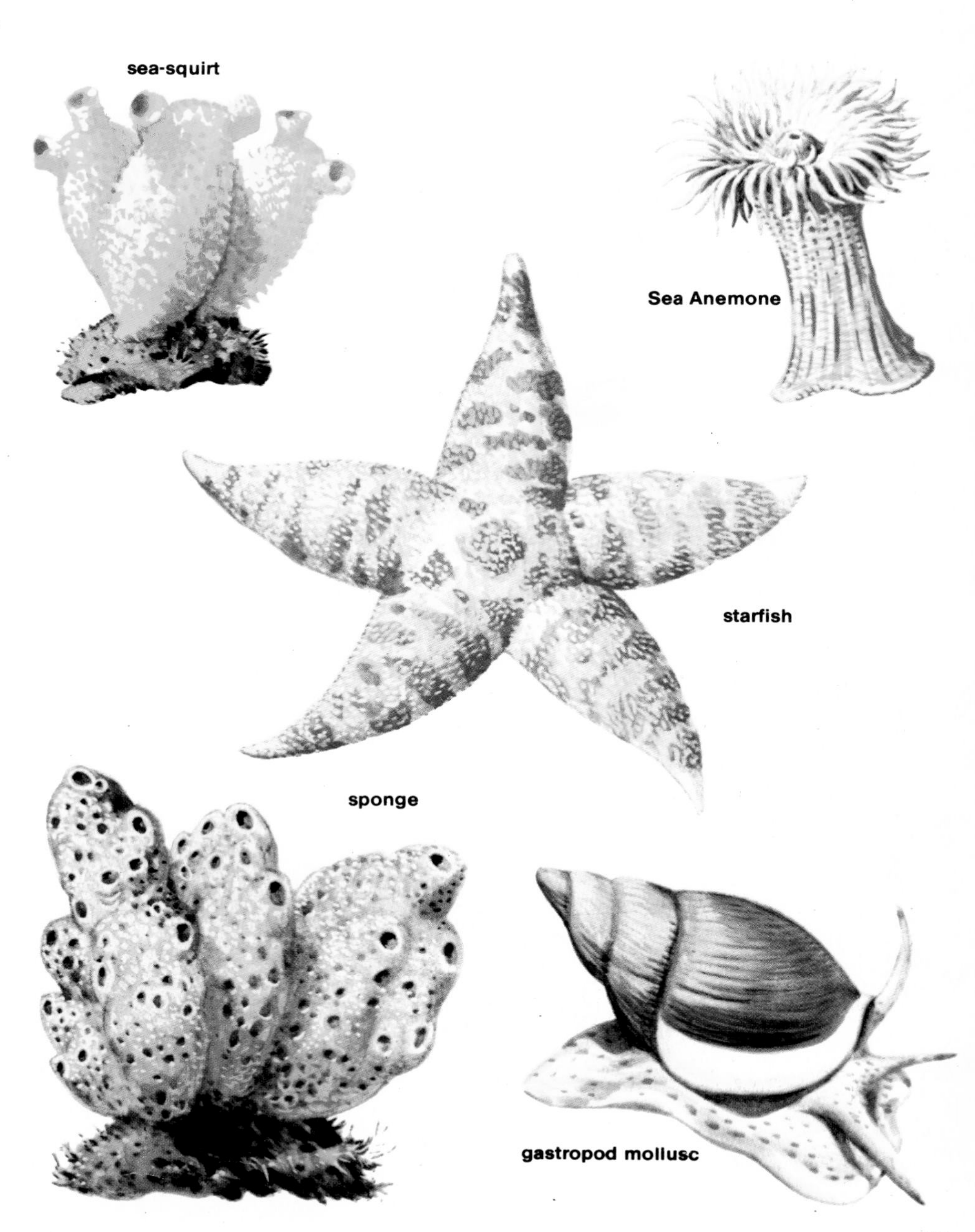

The Biological Cycle of the Antarctic

The Food-chain

The abundance of phytoplankton in the Antarctic creates a plentiful supply of zooplankton. The microscopic animals and larvae of larger marine animals (which make up the zooplankton) feed on the phytoplankton and, in their turn, are the basic food of many animals: molluscs, fish, birds, seals and whales. Whales, in particular, consume enormous quantities of zooplankton, especially the crustacean krill — small shrimps which move in enormous shoals near the surface of the sea — which forms the main part of Whalebone Whales' diet.

Birds in their turn feed on fish, molluscs and cuttlefish. Some seals feed mostly on krill. The Sea Elephant (and the closely allied Elephant Seal) feeds on great quantities of fish and cuttlefish, while the Leopard Seal, the most ferocious seal, eats large numbers of penguins which it attacks both in the water and on land. Dolphins and similar species eat fish, surface-swimming cuttlefish and probably baby seals. The Sperm Whale prefers the giant cuttlefish which inhabit the deepest waters, although they may occasionally take a seal. The voracious and insatiable Killer Whale attacks seals, dolphins, penguins and even other large whales which often bear the scars of wounds inflicted by the sharp teeth of these predators.

The Importance of the Sea

The biological cycle of the Antarctic depends essentially on the sea since the land is virtually sterile. Antarctica can be compared with a desert. Its permanent ice cover, freezing temperatures and strong winds prevent plant development and the existence of the larger forms of animal life on the land.

It is in the sea that the food-chain starts, with the transformation of inorganic matter into organic matter by photosynthesis, and in the sea can be found the final links in the food-chain: the large predators.

ANTARCTIC COAST

SUB-ANTARCTIC – ISLANDS

Antarctic Seals

A typical scene in the Antarctic seas: some Crab Seals are basking in the sun on an ice floe with a group of penguins and skuas.

Life on the Ice

Most Antarctic seals live on ice-floes or on the frozen coastal ice without ever coming on to "dry" land. A typical example is the Weddell's Seal, which grows up to three metres in length. It has dark grey fur on the back and lighter grey on the abdomen. Weddell's Seals spend many hours of the summer basking in the sun on large ice-floes, while in the winter they retreat beneath the ice, breathing through specially made holes.

The Crabeater Seal, another typical Antarctic species, has silvery-fawn fur and feeds exclusively on zooplankton, which it filters through teeth adapted for that purpose. Their numbers are still large and are estimated at be ween three and four million.

The Leopard Seal, an animal which is only found in the Antarctic Ocean is the penguins' greatest enemy. It can be up to 3.5 metres in length and it is slim and gracefully built with a shiny silvery-grey fur. Its mouth is armed with sharp teeth, showing its predatory habits. Though it feeds on shrimps and molluscs it has a preference for penguins, settling itself near their colonies and silently attacking them either in the water or on the ice.

Crabeater Seal
Leopard Seal
Weddell's Seal
Ross Seal
sea-lion
Sea Elephant

Whales of the South Pole

BEARDED WHALES

fin-whale

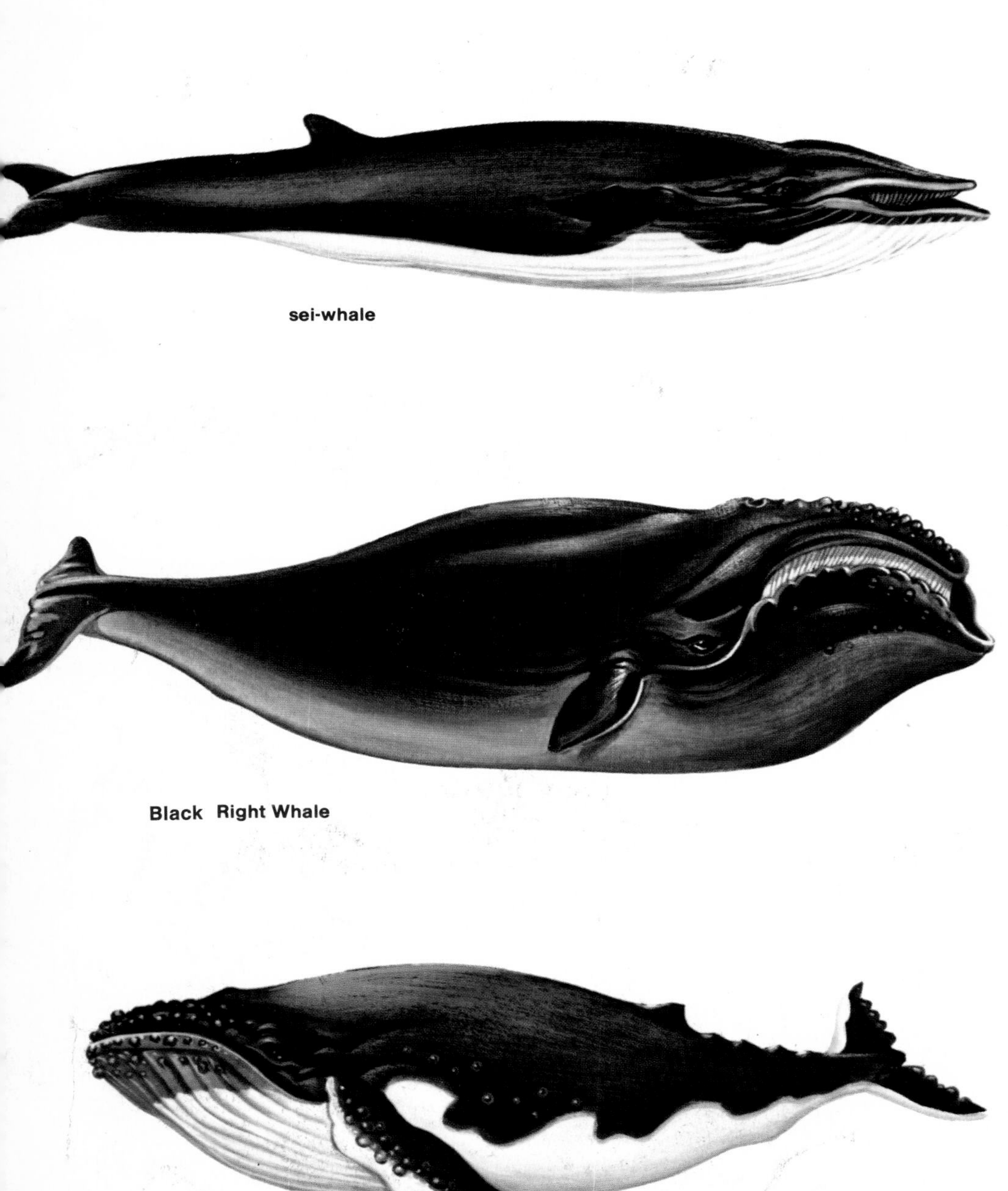

The Search for Food

The Arctic and Antarctic waters are full of krill, the small crustaceans on which some whales feed. The krill swim near the surface, carried along by the currents. Swimming against the current with their mouths open, the Whalebone (or Baleen) Whales have no difficulty in catching their small prey. They swim lazily through the vast shoals of krill straining their food through the whalebone sieves in their mouths. Whales stay in the food-rich waters all through the summer, building up a thick layer of subcutaneous fat.

At the approach of the polar winter, whales start a long migration to tropical waters where the young whales are born. Gestation lasts about one year and the female gives birth to a single pup, which is fairly well-developed and measures about a third of its mother's length. Whales are gregarious and most of them live in schools which increase in size during migration.

Whalebone Whales do not eat during the months they spend in the warmer waters; they depend on the reserves of fat built up during the previous season.

Whales are able to close their blow-holes and store great quantities of air in their huge lungs. This allows them to stay submerged for up to 75 minutes at a time. Sperm Whales can dive to a depth of 1000 metres in

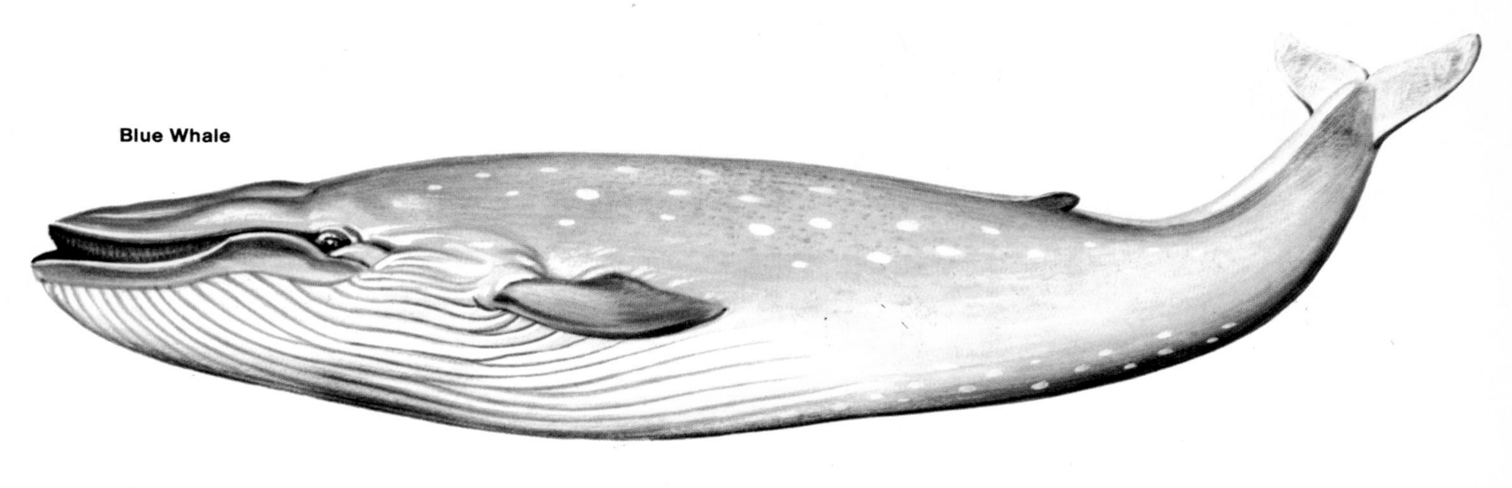
Blue Whale

search of the giant cuttlefish on which they feed.

Killer Whales

In contrast to the peaceful Mysticeti (or Whalebone Whales) the Odontoceti (or Toothed Whales) are aggressive, especially the Killer Whales, whose large mouths contain forty sharp teeth. They swim in small schools and together will attack even larger whales, tearing them to pieces. They kill dolphins and seals with incredible ease; there is a case recorded describing how the remains of twenty seals were found in the stomach of a dead Killer Whale. In both Arctic and Antarctic waters, Killer Whales are distinguished by their two-metre high dorsal fin which is rather like that of the sharks.

Valuable Ambergris

A giant among the Toothed Whales, the Sperm Whale grows up to twenty metres in length and may live in either the northern or southern hemisphere. From the Sperm Whale we get spermaceti, a colourless oil used in the cosmetic industry, in pharmacology and in candle-making.
Sometimes the intestines of Sperm Whales contain a greyish substance called "ambergris" which is highly valued and used in the production of high quality perfumes. Because of the demand for spermaceti and ambergris, Sperm Whales were nearly exterminated at the beginning of this century.

Minke Whale

TOOTHED WHALES

Pilot Whale

Bottle-nosed Whales

White-sided Dolphin

Common Dolphin

Killer Whale

Sperm Whale

Antarctic Birds

At least forty different species of birds nest in the Antarctic region but only five of these are terrestrial. The sea-birds include seven species of penguin, twenty-four petrels, two cormorants, two skuas and one gull. Amongst the non-marine species are two ducks, two sheathbills and one species of lark. The relative paucity of species is more than compensated by the large numbers of each species. The small number of land-birds is not surprising in view of the lack of food on the frozen continent. The marine birds, however, have an abundance of food in the Antarctic seas. Most of the sea birds nest along the edges of the Antarctic continent; although a few, like the Snow Petrel, fly some distance inland to nest on mountain peaks up to 2,000 metres. The Antarctic cormorants always nest near the sea. They fish by plunging from the surface, pursuing the fish underwater. They tend to fish in the shallower coastal waters but can fish even in deep water. Many species including the Kelp Gull, larger skuas and sheathbills nest on the Antarctic Peninsula, which reaches further northwards away from the South Pole and where the climate is less severe and the seas remain free of ice for longer periods of the year. In the more northerly parts of South Georgia, Pintail Ducks nest in marshy ground in areas of fairly rich but low-growing vegetation. On the Kerguelen Islands both the Pintail and another related species nest. Both the ducks feed on small crustaceans and molluscs.

Above: *Terns (or Sea Swallows) are graceful sea-birds with forked tails and silvery grey plumage and usually a black head. They feed on fish.*

Below: *The Giant Petrel (or Giant Fulmar) is a large sea-bird some 76 centimetres long from beak to tail. It is a skilled flier with great gliding ability and is a voracious scavenger. It will feed on any dead animal, attack wounded birds or feed on the eggs and young of other birds. Petrels have an unpleasant habit of squirting foul-smelling stomach oil from their beaks when alarmed. There are two colour phases — light and dark plumage — shown in the photographs below. The darker phase is more common in the more northern parts of their range, while the white phase is common in the south.*

Migratory Birds

When the ice gets thinner during the summer, many birds which nest further north come into the rich Antarctic waters attracted by the abundant food. Amongst these are the Sooty Albatross and the Skuas. Some of the migrants travel long distances, the supreme trans-Equatorial migrant being the Arctic Tern which travels 15,000 kilometres with ease from the northern to the southern polar region. The Sooty Shearwater also crosses the Equator to the northern hemisphere.

Above: *Related to the Giant Petrel, the Antarctic Fulmar (or Cape Dove) nests in colonies on cliffs or in more open areas. They have similar habits to the Fulmar Petrel of the northern hemisphere. Their eggs are incubated by both parent birds and hatch in about forty-five days.*

Below left: *Sheathbills look similar to white pigeons. They are the only birds without webbed feet to nest on the Antarctic continent. They are great scavengers and are common near whaling-stations or anywhere else where they can find offal. They are very tame birds which nest on the ground, rearing one brood a year.*

Below right: *Skuas are large birds, related to gulls, but often with dark plumage. There are two species of skua living in the Antarctic. They are the McCormick Great Skua and Lonnbergs Great Skua. The former lives on the Antarctic continent itself and on nearby islands while the latter species, which is larger and darker, nests on the Antarctic and sub-Antarctic islands. Both species return to their nesting-place after spending a period between May and September in more northerly waters of the sub-tropical oceans. They are aggressive birds by nature, often nesting near penguin colonies on whose eggs and young they feed. They harry the parent birds to make them leave their young which they then seize in one single swoop. The skuas are great aerial pirates.*

The Penguins

Adelie Penguins are excellent swimmers. They winter on the ice-floes but move inland during the mating season. The males are the first to arrive and a month later they are joined by the females. After the birds have selected their mates and courtship has been completed, they begin to build their nests. The nest is a simple structure of pebbles and bones. The female lays two eggs at an interval of a few days, and then leaves the male to incubate them for the first ten days, while she returns to the sea to feed. For the entire period while they are looking after the young the parents take turns both in incubating the eggs and in feeding the young.

Colonies of penguins, tens of thousands strong, live in the Antarctic and the sub-Antarctic. The penguin's body is ideally adapted to sea life. The wings have been transformed into fins, the feet are webbed, and the feathers smooth and thick at the base helping to keep the bird warm. A thick layer of fat adds extra warmth and also serves as a food reserve.
Penguins feed on small crustaceans and fish; larger species catch bigger fish and cuttlefish for which they will dive into quite deep water.
All species, with the exception of the Emperor and Royal Penguins, build nests and deposit two eggs which are then incubated by both parents. Penguins only stay on land during the hatching and raising of their young. The rest of the time they live in the water and very little is known about their habits during that period.

The Emperor Penguin

The Emperor Penguin is a giant bird. Standing erect it is over a metre tall and weighs between 20-46 kilograms.
At breeding time, the Emperor Penguins leave the sea and may travel great distances to reach their nesting-places. Generally they choose the same mate as in previous years. When the egg is deposited the male takes it upon himself to incubate it, holding it between the feet and an abdominal fold, while the female returns to the sea to replenish her food supplies, staying for about two months and often travelling more than a hundred kilometres. By the time she returns, the young bird has been hatched by the male and she takes over the care of the chick. The exhausted male, very much thinner after his enforced starvation, then returns to the sea to feed for four or five weeks. When the young are able to swim the entire colony walks to the sea.

This entire colony of Adelie Penguins left the Ross Ice Barrier, after being disturbed by scientists and their equipment. They regrouped on a deserted beach where they once more established their colony and carried on with raising their young undisturbed.

Rockhoppers, living in sub-Antarctic and temperate regions, nest on rough ground and on steep, sharp rocks overhanging the sea. They manage to climb to their nests with the aid of their curved claws. They weigh between two and four kilograms and, in spite of their small size, they are quite aggressive birds.

Below: *Similar in appearance to the Emperor Penguin, the King Penguin is about 80 centimetres high and weighs between 15 and 20 kilograms. The main difference between them is that the King Penguin is slightly smaller and has a small orange-yellow patch on the neck. The King Penguin lives further north than the Emperor Penguin and forms enormous colonies. After the female has laid her single egg, the male incubates it first, between the feet and the abdominal fold. Later they both take it in turns for the fifteen-day incubation. The young huddle very close together for protection against the cold as they wait for their parents to return and feed them at intervals which may be as long as two or three weeks.*

The Sea Elephant

Gigantic Seals

The Sea Elephant lives only in the southern hemisphere and breeds only on the sub-Antarctic islands, although a related species — the Elephant Seal — occurs in the northern Pacific. It is the largest of the seals: an adult male can measure 6.5 metres in length and weigh up to 4 tonnes. The females are smaller and are usually not more than 3.5 metres in length. They have a layer of subcutaneous fat which, in the males, may be 15 centimetres thick. An oil which is particularly well suited for lubrication is extracted from this fat and for this reason Sea Elephants have been hunted for centuries and have been close to extinction. The male has a kind of trunk with nostrils below it: during the mating season this organ enlarges and acts as a sound box amplifying the mating calls.

Sea Elephants spend most of their lives in the water, where they also spend the winter months. During September — the start of the southern hemisphere Spring — the males return to the sub-Antarctic and South American beaches. They engage in ferocious battles over the possession of territory, attacking one another with sharp, pointed teeth but not actually killing their opponents, even though they may inflict some savage wounds.

In the breeding season, the male Sea Elephants are the first to arrive on the beaches. Once there they fight for the largest share of the beach. A month later they are joined by the females and the harems are formed. Each male has a group of between twenty and forty females which it guards jealously. It even goes without food during the whole of the breeding season.

A month later the females arrive and the males, which by now have their own territories, try to win their affection. The "harems" usually consist of between twenty and forty females but in some cases up to 150 females have been found with one bull Sea Elephant. The male guards

The Sea Elephant lives only in the sub-Antarctic regions and is the largest member of the seal family. The males are up to 6.5 metres long while the females are a good deal smaller. The short trunk, characteristic of the male, appears at the end of the third year of life. It swells up and becomes larger durina the breeding-season.

the females jealously, refusing even to eat. When an intruder comes into the territory — often a youngster without a mate — the bull attacks it fiercely, driving it away. All this time the male goes without food and, as a result, loses a lot of weight during the breeding season.

Colossal Babies

At the beginning of October the young are born. They are about 1.2 metres long and they weigh about 35 kilograms. They are fed continuously by their mothers and they treble their weight during the first three weeks of their lives. The mothers do not eat during this period. In their fifth week the young moult, changing their fur, and enter water for the first time in their lives. The life of a young Sea Elephant is not without danger. In their first weeks they are prone to sickness, but more often become victims of Killer Whales and Leopard Seals. Only the fortunate and the fit will survive to live in the rigorous climate.

In spite of their enormous size, Sea Elephants are extremely agile in water. Their spines are very flexible and on land they are able to touch the extremities of their bodies with their trunks. This does not, however, help them to walk on land where these creatures, so graceful in the sea, waddle awkwardly.

The Albatross

Above: *The strong, slim long wings of the albatross allow it to fly for hours using air currents. These birds glide for great distances reducing muscular effort to a minimum. The albatrosses which breed in the Antarctic may fly up in the non-breeding season to the northern hemisphere.*

A Life at Sea

Essentially a sea-bird, the albatross spends the greater part of its life flying effortlessly over the sea. Out of thirteen species of albatross, nine nest in the southern hemisphere. Among them the Wandering Albatross, a giant among sea-birds, which nests in South Georgia, on the Kerguelen Islands and on some other small islands of the sub-Antarctic. This huge bird weighs about 10 kilograms with a wingspan of up to three metres. The larger species of albatross lay only a single egg every two years. The young are fed for four or five weeks by the parents but, as winter approaches, the parents leave them for long periods, returning to the sea to search for food. The fledglings put on weight rapidly and by the following Spring they are ready to leave the nest, having spent nearly 300 days there. Even the smaller species of albatross has a wingspan of two metres.

Below: *Albatrosses usually mate for life, which can be up to fifty years. Larger species reproduce only every two years, the mating being preceded by an elaborate courtship display which lasts two or three weeks, during which time the nest is built.*

Above: *The Grey-headed Albatross, smaller than the Wandering Albatross but still with a wingspan over two metres, nests in colonies high up on the rocks. The nest has a cylindrical shape, about half a metre high, constructed of grass reinforced with mud. The egg is deposited on the hollowed top.*

Below: *The Sooty Albatross is a relatively small bird with a graceful body covered in dark plumage. Pairs nest apart from other birds on the sharp rocks overhanging the sea on South Georgia and other Antarctic islands, coming in as far as the southern ice cap. The fledgling, born on a cliff, starts its flying life when it makes its first glide off the nest.*

Below: *An adult Black-browed Albatross. When they are a month old, the chicks of the Grey-headed and Black-browed Albatrosses are left alone while the parents go in search of food. The young of these species have a special means of defence: if alarmed they eject a spray of evil-smelling oil which helps to repel the attackers.*

The Antarctic Base Camps

The first Antarctic base was planned and organised by an American, Richard E. Byrd, and was named "Little America". Situated near the Bay of Whales, from where Amundsen set off for his successful expedition to the Pole, Little America was used as a base by the many scientific expeditions — using aeroplanes, ships, aircraft-carriers and ice-breakers — which gathered valuable geographical and scientific data about the Antarctic.

During a conference in 1934 it was decided that the Antarctic should be divided between Great Britain, New Zealand, Norway, Australia and France. This decision was not accepted by all nations, particularly the U.S.A. and U.S.S.R., and for a long time there was international controversy about the continent. The establishment of the International Geophysical Year (1957-1958) saw the start of co-operation between the nations which claimed sovereignty over the Antarctic continent. As a result forty-two permanent stations were established where eight hundred technicians are permanently employed: the Americans installed a station on the South Pole itself, named "Deep Freeze". Scientists and technicians of different nationalities have made use of purpose-built machines (such as "snow-cats" and similar vehicles) as well as high-precision scientific instruments capable of resisting the bitter polar climate. With these intruments it has been possible, among other things, to determine the thickness of the ice, to obtain data about temperature and to study solar radiation.

In spite of having modern machines for transport, sledge-dogs are still used in the Antarctic research stations.

Below: *Little America was the first base camp to be established on the Antarctic continent. Planned and organised by an American, Richard E. Byrd, it is near the Bay of Whales and its supplies arrive by icebreaker, aircraft and helicopter.*

Above: *Ice-breakers manage to force their way through the frozen pack-ice to open the way for other shipping. Ice-breakers which supply the American station at McMurdo come within 1.500 kilometres of the South Pole.*

Above: *Caterpillar-tracked vehicles have proved indispensible in the exploration of the Antarctic. They are often used for the transport of scientific equipment on expeditions into the interior of the continent.*

Above: *The introduction of aircraft has given an impetus to the exploration of the Antarctic. In order to facilitate landings on icy runways, the planes are fitted with special skid plates. The use of helicopters by scientists has meant that the small islands, which were previously inaccessible, can now be studied.*

Whaling

Whaling first started in Arctic waters. After the slaughter of 1,500 whales, the numbers of whales in the northern waters were considerably reduced as whales are very slow breeders. The hunters then turned their attention to the Antarctic seas. It was in fact the Englishman, Captain Ross, who first proved that there was an abundance of whales in these seas. Many British and Norwegian firms realised the enormous wealth that could be made from such hunting. It is possible to extract about twenty tonnes of good quality oil from a single whale. This could be used either for cooking or as

Above: *With the use of strong lifting gear the whale is hauled aboard the factory-ship onto a special deck which tilts down to the level of the sea. The whale is then carefully cut up and the pieces sorted, because each part of the body has a particular use.*

Below: *A dead whale is fixed to the whaler by its tail and pumped full of air until it floats. The whaler then returns with it to the factory-ship where the whale is transferred and firmly attached to the ship.*

combustible oil. Whalebones were very valuable as they provided the only strong and flexible substance that was available to Man in the nineteenth century.

Before steamships and explosive harpoons were used, whaling was carried out from rowing-boats by men armed with harpoons. The light vessels approached dangerously near to the enormous mammals in order to strike, at great risk of being overturned.

Moby Dick and Captain Ahab

Herman Melville — an American writer of the last century — gave us a precise but gripping description of

whaling in his famous novel *Moby Dick*, written in 1881. The story tells of Captain Ahab who for years hunted a particular White Whale which he called Moby Dick and which finally killed him, dragging him down into the depths.

Above: *Jets of steam and sprays of water are sure signs that a school of whales is near. Nowadays, the introduction of radar has completely revolutionized the technique of whale-sighting and they can be detected much more quickly.*

Below: *It sometimes happens that a school of whales, which seems to have become disorientated, gets stranded in shallow coastal waters. As the water goes down the whales die, suffocated by their own weight when out of water.*

Explosive Harpoons

During the first decades of this century, with the introduction of steam-ships and explosive harpoons, hundreds of thousands of whales were killed and some species, particularly large ones, were almost exterminated. It was only in 1962 that effective regulations were introduced which limited whaling. It had not been possible to make any such international agreements previously because of the interests of the different countries — their financial involvement was great and nobody wanted to reduce the catch.

Animals of the Polar Regions

Arctic Fox

Adelie Penguins see Penguins

Arctic Fox *(Canis lagopus)* Arctic

Only slightly smaller than the Red Fox. The coat is soft and thick and changes from browny-grey in the summer to snow-white in winter. The Arctic Fox lives in the Arctic regions of Europe, Asia and America up to 60 degrees of latitude. They are not frightened of Man and behave almost like dogs. It is not rare to see them in Eskimo villages where even during daylight they search for meat waste. They are hunted for their fur.

Arctic Tern *(Sterna paradisaea)*

Arctic Wolf see Wolves

Auks *(Alcidae)*, Arctic

A family of Arctic sea birds, 13 species. They breed in the sub-Arctic and moderate regions, then move to the very north. Auks have a plump body, narrow pointed wings, a short tail and thick bill. They feed on fish, crab and plankton from the surface – although some do dive. Nesting colonies on sea cliffs. They lay only one egg and both parent birds sit on the nest. Most common species are the razorbill, puffin and guillemot. There used to be the Giant Auk, which could not fly, but the last two individuals were killed on Iceland in 1844.

Giant Auk

Banded Seal and Bearded Seal see Common Seal

Beluga see Whales

Caribou see Reindeer

Common Seals *(Phocidae)*

A family with many subspecies. They are not able to put their back flippers underneath their body and thus find it difficult to move on land. The most important species are:

Banded Seal *(Phoca fasciata)*

Up to 1.7 metres long, chocolate brown coat which is covered in wide yellow or whitish bands. They live in the drift-ice regions of the Bering Sea and in the Ochotic Ocean.

Bearded Seal *(Erignathus barbadus)*

This is the largest seal and the male can grow up to 3 metres long. They have long moustaches, the coat is grey to yellowish-brown on the back and yellowish to silver-grey underneath. Mostly around the North Pole, although some will stray to the European coasts.

Razorbill

Common or Harbour Seal *(Phoca vitulina)*

For us this is the most common seal. Because the head is similar to that of a dog, it is also known as the Dog Seal. It is about 1.8 metres long with light to dark grey coat. Occurs in areas around the north coast of Europe and Canada and reaches as far as the North Sea. There are occasional sightings on the Baltic coast, and sometimes young get lost too far inland. If found, the nearest zoo should be contacted.

Harp or Greenland Seal

Harp or Greenland Seal *(Pagophilus greenlandicus)*

Up to 2.0 metres long, grey to yellowish coat with black horseshoe-shaped band across the back which resembles a saddle. Lives on drift-ice in the Arctic Atlantic. There are herds which will always return to the same place for breeding and rearing the young: at the St. Lawrence Gulf, the coasts of Labrador, Newfoundland and the seas around Greenland.

Ringed Seal *(Phoca hispida)*

Up to 1.6 metres long, often with dark grey coat and black spots, many of which are ringed with pale hair. They live in Arctic waters but can reach as far as the Baltic Sea.

Ringed Seal

The so-called South Seals which live only in the Antarctic are divided into four subspecies:

Crabeater Seal *(Lobodon carcinophagus)*

Feeds mainly on plankton. It reaches a length of 2.6 metres and has a silver-grey coat with chocolate brown spots on shoulders and flanks; in the summer this changes to all over yellowish-white. Lives on the drift-ice of the Antarctic, the south coast of the Argentine, Australia and New Zealand.

Leopard Seal *(Hydrurga leptonyx)*

The patterns on the coat are reminiscent of that of a leopard. The male is brown with yellow patches. They are up to 4.0 metres long and have the powerful teeth of a beast of prey. The Leopard Seal feeds on young seals and penguins, but also on waste from whaling vessels. They live in the Antarctic, the Falkland Islands and Tasmania.

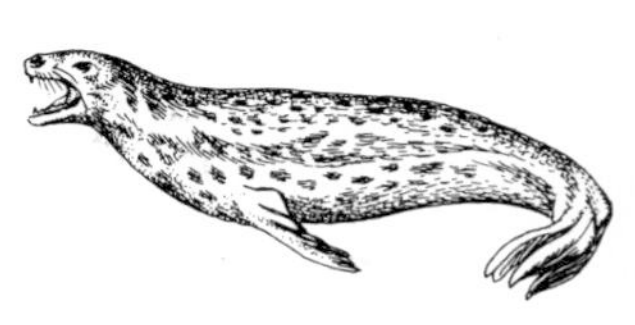

Leopard Seal

Ross Seal *(Ommatophoca rossi)*

There are about 20,000 left of this kind. On the top, the coat is greyish-yellow with yellow stripes on the flanks; it is paler on the belly. They live on the central drift-ice around the Antarctic. Their eyes are particularly large.

Weddell's Seal *(Leptonychotes weddelli)*

Up to 3.0 metres long, ice-grey coat with yellow-white patches, lighter on the underside. During the winter they live underneath the polar pack-ice, in which they make breathing-holes. Will also come to the south coast of Patagonia and the southern Orkney and Falkland Islands.

The largest of all seals is the:

Elephant Seal *(Mirounga)*

Old bulls will grow to 9.0 metres in length. They live only on the southern half of the globe, mostly along the south coasts of South America. Despite their size they are very agile in the water, but clumsy on land.

Only the bulls have the trunk; when the animal is agitated, the trunk swells up. Northern and Southern Elephant Seals are recognized. Only the latter will come to the waters of the Antarctic.

Crabeater Seal see Common Seals

Divers *(Gavia)* Arctic

This bird family has four subspecies, which all breed in the polar regions. They are between 58 and 90 centimetres long and have a straight, strong bill, short wings, large webbed feet and very thick feathers. They nest along the banks of lakes and in swamps. Generally there are two eggs, spotted brown. Being good swimmers and divers, they catch their prey from a 40 to 50 centimetres depth. The largest species is the **White-Billed Diver** *(Gavia adamsii)* followed by the **Great Northern Diver** *(Gavia immer)*, and the **Blackthroated Diver** *(Gavia arctica)*. The smallest is the **Red-throated Diver** *(Gavia stellata)* whose wings are somewhat longer, so that he needs a shorter starting run than the others and can thus live on smaller lakes.

Dominican Gull see Gulls

Eared Seals *(Otariidae)*

They live in the cold regions of the globe, mostly around the North Pole. They have a visible external ear. All Eared Seals are able to move their rear flippers underneath their body, and lift up their front flippers. This enables them to walk quite fast.

South American Sea-Lion *(Otaria byronia)*

They have long, silken hair, which will fall like a mane with the male animals. This seal will reach a length of 2.45 metres, and colours can be from dark-brown, to grey to golden yellow. They are found chiefly on the southern coast of South America and the Falkland Islands. These sea-lions are very intelligent and easily trainable.

Steller's Sea-Lion *(Eumetopias jubata)*

This species is the largest of all eared seals. The male may grow up to 3.5 metres in length. His coat is hard, coarse and yellowy-brown. Steller's Sea-Lion lives in the North Pacific and the Bering Sea. During the time of mating he moves in large groups to the coast of Alaska, the Pribilof Islands and the Aleutian Islands.

Fur Seals, also called Bear Seals, are the true Fur Seals. Underneath the coarse top hair they have thick woolly fur, which is why Man constantly hunts them. Their pelt is priceless.

Alaska Fur Seal *(Callorhimus ursinus)*

They live in the North Pacific and the Bering Sea. The male reaches a length of 2.5 metres. In Spring, when the mating-season starts, they move to the coasts of Alaska and the Pribilof Islands. Here, ferocious fights take place between the rival males. Because of the silvery fur they have been much hunted, but now they belong to the protected species. In 1870, there were roughly 4.5 million; in 1914 only 200,000 could be found. Since the protection law they have started to increase again.

Alaska Fur Seal

Southern Fur Seals and Sea-Lions *(Arctocephalus)*

There are seven different subspecies which are not the same in size but have the same kind of fur: mostly black, mixed with white. This fur is extremely sought after and the animals are thus hunted everywhere. The behaviour pattern is similar to that of the northern species. Bulls await the females in the Spring in certain places along the coast. After the young (which were conceived the previous year) are born, the females are ready for mating again.

Elks *(Alcinae)* Arctic

The largest of all deer. Massive body with height of 1.8 metres to 2.0 metres and a weight of 500 kilograms. Elks have a huge head with a large, freely moving upper lip, which they use to tear off food. The legs are long and powerful with large hoofs and wide spreading toes, which enable them to walk safely on snow and mud. Their fur is grey to yellowish. A characteristic is the huge, shovel-like antlers of the male. Both sexes have a clump of tissue and hair under the chin which is 20 to 25 centimetres long. Elks live in herds and can run and swim well. Two kinds are distinguished: the European Elk living in the forests and tundras of northern Europe and northern Siberia – these have lately been successfully tamed in the USSR, to pull and carry loads and as dairy herds. The other is the North American Elk, called moose, which is slightly larger and lives in the northern forests of Canada and Alaska.

Elk

Emperor Penguin see Penguins

Fulmar *(Fulmar glacialis)* Arctic

This sea bird which is related to the gull, lives in the icy oceans of the Arctic and nests in Greenland, on Novaya Zemlya, on the Kamschatka Peninsula and the Aleutian Islands. Fulmars, on the whole, remain near their nesting-places. Only one species, the Atlantic Petrel, has recently come as far as southern England. Presumably, they have followed the big whaling-ships. The fulmar is 50 centimetres long and has large pointed wings. Feathers are grey-blue on the back; head and underside are white. Lives on the open sea and flies even during strong gales. On land they move awkwardly. They feed on crabs, fish and molluscs.

Fulmar

Giant Fulmar *(Macronetes giganteus)* Arctic

They are at home in the Antarctic and the South Pole. This fulmar is the largest one and reaches a length of 90 centimetres. He has a long hooked beak, a strong, thick rump, wide pointed wings and short legs with webbed feet. Feathers are brown with white edges. The Giant Fulmars live on the open sea and are very strong swimmers. They feed on fish, but will also take waste from ships, young sea birds and penguin eggs and young.

Fulmars and shearwaters *(Procellariidae)*

These sea birds belong, like the albatross, to the family of tube-nosed birds. The long nose-tube protects them from penetrating water while diving. They have large nose glands which will reject excess salt.

Giant Fulmar

Fur Seals see Eared Seals

Geese

Canada Goose *(Branta canadensis)* Arctic

This goose can grow to more than 100 centimetres in length and is the largest of the true geese. We find them in the whole of North America but also in England, Sweden and New Zealand. Has been turned into domestic goose; living in the wild, their numbers have grown to such an extent that farmers have to protect their meadows. A smaller relative, the Barnacle Goose (70 centimetres), and the Brent Goose (60 centimetres) live further in the north, in Greenland and in Siberia.

Snow Goose *(Anser caerulescens)* Arctic

Most of the Snow Geese have a pure white plumage with black wingtips, but there are also subspecies which are grey. Snow Geese measure about 90 centimetres in length, and live during the

Canada Goose

Golden Plover

Guillemot

Glaucous Gull

Skua

summer around the Hudson Bay up to the Aleutian Islands. There are smaller numbers in the Arctic regions of Eurasia. They migrate for the winter to warmer countries.

Glaucous Gull see Gulls

Golden Plover *(Pluviales apricaria)* Arctic
Largest and most magnificent bird from the family of plovers, reaching a length of 26 centimetres. The name is derived from the numerous golden-yellow spots on the dark brown back feathers. He lives and nests during summer in the north of Europe and Asia, and covers up to 10,000 kilometres to spend the winter in Australia and New Zealand.

Great Northern Diver see Divers

Guillemot or murre *(Uria)*
Guillemots are the largest of all auks and are sometimes called the "penguins of the north". They breed in huge colonies. They can also be found along the coasts of Heligoland, where they breed in their thousands.

Gulls and terns *(Laridae)*
Gulls are found all over the world. They live in large colonies along the sea-coasts, lakes and rivers. They are strongly built but slim, with a strong bill, long pointed wings and webbed front toes. Gulls are superb flyers and swimmers. They feed, depending on the species, on fish, mussels, crabs, small land animals and waste.
Dominican Gull *(Larus dominicanus)*
They occur in the Antarctic and belong to the large gulls. They grow to about 58 centimetres long and live in the southern regions.
Glaucus Gull *(Larus hyperboreus)* Arctic
Up to 70 centimetres long, with wax-white plumage. They breed mostly in single pairs. Northern seas of Europe and Asia and the Americas are their home, but during the winter they migrate to the Mediterranean, California and Florida.
Herring Gull *(Larus argentatus argentatus)* Arctic
Up to 60 centimetres long, with white-grey plumage, dark stripes at head and neck, black wing tips. They feed on crabs, worms, and molluscs which they drop on to stones in order to break them open. They live in the northern areas of Europe and the Americas; in the autumn they move to the French coasts or the Mediterranean.
Iceland Gull *(Larus argentatus glaucoides)* Arctic
Up to 55 centimetres long, the plumage is similar to that of the glaucous gull. During the summer they live along the coasts of Iceland and Greenland. During the winter they move to the north-east coasts of Europe.
Kittiwake *(Rissa tridactyla)* Arctic
They nest in large colonies on coastal rock faces. Along the steep coasts of Greenland there are colonies of about 100,000 pairs and more. They will also come to the British Isles, Norway, Iceland and Heligoland. Only at breeding-time do they come ashore, being the only true sea-gull. Male and female together build their nest from water-plants which have been mixed with mud, constructing a cup-shaped nest on the cliff.
Ross' Gull *(Rhodostethia rosea)* Arctic
One of the rarest gulls, the only one with a wedged tail; white or pale grey in colour. During the summer they appear in the very north of Eurasia and North America. The only breeding-grounds are in the Kolyma Delta of North Siberia.
Skua *(Stercorarius)* Arctic
There are four subspecies in the family. The largest is the Great Skua. They are migrants. Skuas breed in both north and south polar regions. Most of them along the cliffs of the Antarctic coasts. During the Antarctic winter they move north, and the Skuas of the Arctic move south. They are very often seen robbing other birds of fish. Great Skuas are extremely aggressive and steal eggs and chicks.

Humpback Whale see Whales

King Penguin see Penguins

Lemming *(Lemmus)* Arctic
They live in Siberia, Scandinavia, Canada and Greenland. Lemmings have a thick head and plump body. They are about 15 centimetres long, with short legs and sharp claws on their feet. The coat is thick, brownish or grey in colour and some species turn white in winter. They move at night and feed on grass, moss and seeds. Some years see large numbers of young and that results in overpopulation. That is the time when they leave their environment in enormous numbers, pushing each other as they rush along. Many of them die in rivers, lakes or in the ocean.

Leopard Seal see Common Seals

Musk Ox *(Ovibus moschatus)* Arctic
This ox grows to about 1.2 metres shoulder height; and, without the tail, 2.3 metres long. The body is stumpy, with short legs and a large head. The longish horns curve down first and then up. The Musk Ox's hair is long and thick – the longest hair of any wild living animal. These strange animals used to be at home in Asia, Europe and North America. Now they live only in Greenland and on the Hudson Bay. They get their name from a strong musky smell which they have during the mating-season. Musk Oxen are protected animals.

Narwhal see Whales

Parry Suslik see Susliks

Penguins *(Spheniscidae)* Arctic
Arctic family of sea birds with thirteen subspecies, which have adapted in their own way to life in the water. They cannot fly; their wings have become flippers. This makes them very different from other sea birds. Size can be anything from 40 to 123 centimetres high and they have a long elongated body. Legs are very far back and used as a rudder. The body is evenly covered with feathers. Penguins have a thick layer of fat for insulation. They are very good swimmers, but also move well on land. Their walk can look quite amusing because they either slide on their stomachs or walk upright. Rockhoppers jump with both feet at once. They are mainly found in the Antarctic, but also along the coasts of Australia, New Zealand and South Africa – one species even on the Galapagos Islands. They live only on land during the breeding-season, when they breed in very large colonies. Emperor, King and Adelie Penguins all breed in the Antarctic.

Polar Bear *(Ursus maritimus)* Arctic
Lives in the Arctic regions of Europe, Asia and the Americas. Polar Bears hunt for meat and fish and have totally adapted to life on the ice. They are up to 1.5 metres tall and are covered with a thick, soft white or yellowish fur coat. Their movement on the ice is fast and long distances can be covered easily. Favourite food consists of seals, dolphins, fish and small seabirds.

Ptarmigan *(Lagopus mutus)* Arctic
It lives in Iceland and Greenland, but occurs also in the Alps and Pyrenees.

Lemming

Musk Ox

Emperor Penguins

Polar Bear

Raven

Rednecked Phalarope

Reindeer

Caribou

Sheathbill

Summer plumage brownish with pale grey spots and stripes; winter plumage white. The environment they live in is impossible for any other chicken-like birds. The ptarmigan however goes underground where it finds leafbuds and shoots under the snow.

Raven *(Corvus corax)* Arctic

This bird is found all over the Arctic regions. The largest species of raven is up to 70 centimetres long. They are very strong and the black plumage has a reddish sheen. Different from the black crows in having a wedged tail and a stronger, more curved bill. Often performs acrobatics in flight. They live in pairs or small groups. Feed on worms, insects, slugs, small rodents, fox cubs, smaller birds and various plant matter. The call is very deep and hoarse. This hoarse call has brought them the name "Prophets of Approaching Doom".

Razorbill see Auks

Rednecked Phalarope *(Phalaropus lobatus)* Arctic

They are one of three species of phalarope. They are equally good swimmers as they are runners. On the water they float like cork, nod their heads constantly and move around in circles in order to wash up food from the shallow mudbed below. While the Grey Phalarope is totally red on the underside, the Rednecked Phalarope only has red on his neck.

Reindeer *(Rangifer tarandus)* Arctic

They are different from all other deer in that both male and female have antlers. Reindeer live in the furthest north and along the coasts of the large islands around the North Pole. There is only one kind of reindeer with twenty subspecies, three of which have already died out; six others are endangered. North American reindeer are called caribou. This name comes from the Red Indian language and means "snow mover". Indeed, all wild-living reindeer find their food – moss, grasses and roots – underneath the snow. The caribou are larger than the reindeer of the Old World. The reindeer which live in the forests are larger and have a darker coat. They have large hoofs with spread-out toes, so that they do not sink down in the snow or soft tundra. Only a few wild reindeer are left in Europe.

North European Reindeer

They are 1.8 metres long and 1.0 metres high and have long since been domesticated. They are used as work animals and are bred for their meat. Milk, butter and cheese also come from the reindeer herds; and the skins are used for clothing and tents. Caribou cannot be tamed. They are well-known for their migrations. The so-called Barren Ground Caribou will travel 500 kilometres south in the autumn and return to the tundra in Spring. The West Canadian Waldren is the largest caribou. It is a very strong deer with a body length of 2.2 metres and a height of 1.5 metres.

Seals *(Pinnipedia)*

They are meat-eating mammals, living in the oceans. Their feet have turned into flippers. Seals are amphibious animals. They breed and change their coat on land. They spend most of the time in the water. They are good swimmers and feed on fish, crustaceans and molluscs. The subfamilies are Common Seal, Eared Seal and walrus.

Sheathbill *(Chionis)* Antarctic

These birds are sometimes called Antarctic Pigeons. They have a strong bill and short legs. Because of the fat insulation under the skin they can live well in the Antarctic winter. Their wings are short and the feathers completely white. There are two species: White-faced and Black-faced sheathbills. They live in large numbers on the coasts of the Antarctic islands and South America. They feed on small fish and penguin eggs. Mostly they stay on the drift-ice and in shallow water, away from the open sea.

Silver Gull see Gulls

Snipe *(Gallinago)*

Snipes can be found anywhere on the northern half of the globe. They breed normally in marshes, meadows and swamps; many breed on the Arctic tundra. Most snipes migrate to warmer countries, but some remain over winter in their breeding area, such as England.

Snow Bunting *(Plectophenax nivalis)*

A small bird belonging to the family of finches, about 18 centimetres. Nearly white plumage gives good cover in this environment. During the summer they live on Franz Joseph Land, Novaya Zemlya, Spitzbergen and other northern islands.

Snow Goose see Geese

Snow Hare *(Lepus americanus)*

Lives in the north of Canada and Alaska. Smaller than the Varying Hare. Hind-feet are large and covered with hair; they are used like snowshoes. Summer coat is brown. Winter coat is white. These hares breed very irregularly but there is no known reason for this.

Snowy Owl *(Nyctea scandiaca)* Arctic

With a length of 70 centimetres they are nearly as large as the Eagle Owl. The Snowy Owl has dense white feathers which have darker bands across. They live in the Arctic regions of Europe, Asia and America. They fly during the daytime and feed mainly on rodents (lemmings), fish, crustaceans and molluscs. Sometimes they catch Arctic Hare. Snowy Owls live in the open.

Stoat or ermine *(Mustela erminae)* Arctic

They have been hunted for hundreds of years for their precious coat which adorned the robes of kings. They are meat-eaters and belong, as do the weasels and mink, to the family of martens. The stoat lives in northern Europe, Asia and the Americas. The overall length is 40 centimetres, one third of which is tail. The coat is soft, reddish brown in the summer, white with black tail tips in the winter. Stoats are very agile, climbing and jumping. They live in forests up to 3,400 metres high, but really anywhere where food is to be found. Main food sources are rodents such as mice, hamsters, lemmings and rats. Occasionally they will attack rabbits or chickens.

Suslik *(Citellus)* Arctic

This little rodent is a relative of the squirrel, but lives on the ground. The Parry Suslik ranges from Alaska to the Hudson Bay. They are 22 to 35 centimetres long, the tail 8 to 15 centimetres. The fur is yellowish brown to reddish brown with irregular paler spots. They live in loose communities and burrow complicated systems of tunnels.

Sword Whale see Whale

Terns *(Sternidae)*

They are somewhat larger and with a longer bill than that of terns in other regions; more elegant than gulls and recognizable by their forked tail. Terns are a family all by themselves. The largest is the 55 centimetres-long Caspian Tern.

Snipe

Snow Bunting

Snow Hare

Snowy Owl

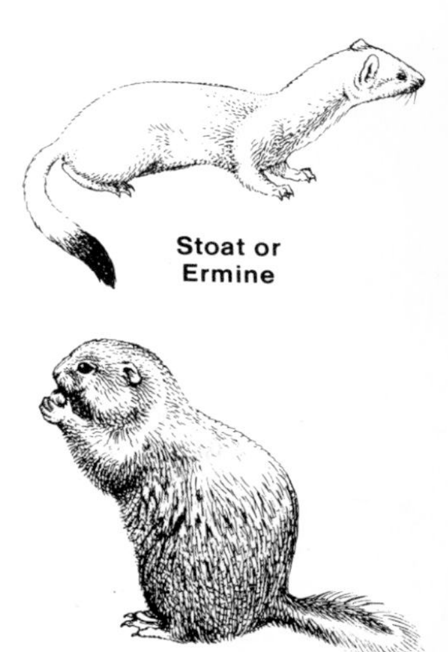
Stoat or Ermine

Parry Suslik

Arctic Tern

The Arctic Tern is 40 centimetres long, with white feathers, dark head and red bill. They are formidable fliers and feed on fish. These terns live in large flocks in the Arctic and breed there. With the onset of winter they cross the Atlantic and head for the Antarctic. The most frequented places are the southern pack-ice regions. They cover up to 30,000 kilometres.

Varying Hare *(Lepus timidus)* Arctic
Grows to about 70 centimetres in length. Lives in the polar regions of the American continent. Of strong build, with small ears. Fur is browny-grey in the summer and pure white in the winter. The tips of the ears are black. These strong animals are not very choosy with their food. In summer they eat herbs and during the winter they eat bark and small branches from willow trees, hazel and other shrubs. The soles of their feet are covered in thick fur, which enables them to move well in the snow. The largest Varying Hare lives in the very north, e.g. Greenland.

Walrus *(Odobenidae)* Arctic
The two upper incisors have developed into a pair of tusks which can be more than 60 centimetres long. For thousands of years Arctic people used to produce carvings from these tusks. Muzzle is blunt with a moustache of stiff bristles – the walrus beard. Size 4.5 metres and nearly one tonne in weight. There are three subspecies which are all endangered. The walrus is plump and massive and has a very thick skin. Only the young are covered in soft fur, which eventually disappears. Range along the coasts of North America and Alaska. Feed on molluscs and crustaceans, which they dislodge with their tusks.

Whales Arctic and Antarctic
Whales are the largest mammals on earth. They live in the sea. They are so heavy that they cannot support their own weight on land. There are Toothed Whales and Bearded Whales. The Bearded or Baleen Whales are so-called from the plates of baleen or whalebone in their mouth. These are used for straining food from the water. Subspecies of the Baleen Whales are the rosquals and the Smooth Whales. Rosquals are: the Blue Whale, 30 metres long, the largest of all whales, the fin-whale, the seiwhales, the Minke Whale and the Humpback Whale.

Whalebone
(from skull of Baleen Whale)

Beluga or White Whale *(Delphinapterus)*
Grows to a length of 5 metres. He has a rounded forehead and slight beak. The young are at first dark grey, then yellowish, and at the age of five years, white. They live along the coasts of Arctic oceans. Sometimes a single beluga will swim up a river.
Greenland Right Whale *(Baleana mysticetus)*
This species was once nearly destroyed. The protection laws should help to increase the numbers again. The Greenland Whale is about 18 metres long, two fifths of that is head. His lower jaw is 5 to 6 metres long and more than 3 metres wide. The tailfin is nearly 2 metres long and about 7 metres wide. They feed on plankton and small crab.
The subspecies of the Toothed Whales are very different from each other. They live mainly on fish.
Humpback Whale *(Megaptera novaeangliae)*
Grows up to 15 metres in length and can be found in all oceans near the coast. He has very long side fins. His baleen is about 60 centimetres long. He is very playful. He jumps out of the water and falls back with a huge splash.

Humpback Whale

Killer Whale or Grampus *(Orcinus orca)*
Killer Whales are the greediest of all whales. They grow to a length of 9 metres and have a large, triangular back fin, which reaches the height of 2 metres with almost all males. They eat fish, but also sea birds, penguins and seals; sometimes even smaller whales. Herds of Killer Whales will sometimes attack Bearded Whales and tear fins and lips off them, so that they bleed to death. There is no proof that a Killer Whale ever killed a human being.
Narwhal *(Monodon monoceras)* Arctic
Distinguished from all other cetaceans by a single spiral tusk in the male, which can grow up to 2.5 metres long. Long ago people believed this to be a horn with magic powers. This is why this "Unicorn of the Sea" used to be hunted. Today he is left in peace, although we still do not know the meaning of the tooth. The narwhal is greyish-blue with dark brown spots.
Pilot Whale *(Globicephala)*
Pilot Whales belong to the dolphin-like whales, although they lack the beak-like mouth. They have a bulbous forehead and slender flippers. Size, up to 3.6 metres to 8.5 metres long. The common Pilot Whale lives in the North Atlantic and feeds mainly on cuttlefish. They form herds of hundreds of animals, and seem to follow the leader whale blindly. If one of the herd gets injured, he roars with pain and races away, only to be followed by the entire herd. That is why they often get washed up and stranded on beaches. In Newfoundland 5,000 Pilot Whales, washed up on the beach, get killed every year.

Pilot Whale

Sperm Whale *(Physeter macrocephalus)*
Maximum length 25 metres; the largest Toothed Whale. Lives on cuttlefish. Has a huge cylindrical head. Some Sperm Whales have in their intestines a very sought-after substance: ambergris, a dark, firm material, which is used for the production of cosmetics. Also have a thick blubber insulation. These are the reasons why they have been so much hunted.

Wilson's Petrel *(Oceanitas oceanicus)* Arctic
Wilson's Petrels are the smallest of all sea birds. There are nineteen kinds with many subspecies. In length between 11 and 22 centimetres. All have a hooked bill and long narrow wings. Feathers dark brown or blackish on the top, whitish on the underside. All petrels feed on the plankton on the water surface, on small fish and molluscs, but will also take waste from ships. Some will dive for food. They are spread all over the world but will breed only in the polar regions. Their flight is elegant and they cross thousands of kilometres to get to their breeding-places. In the Antarctic only the 11 centimetres-long Wilson's Petrel, will breed.

Wolf *(Canis lupus)* Arctic
A few species occur in North East Europe, Siberia and North America. Single ones will come to Scandinavia and the Balkan mountains. The North American Timberwolf has the darkest fur. The largest wolf is at home in the very north, the Arctic Wolf, which has a completely white fur coat. All wolves hunt in packs. One lead-wolf can take the pack over long distances. Feed on larger ungulates, such as reindeer, elk and also domestic animals and smaller mammals. They tend to stay away from people, although they are very much feared and have been hunted nearly to extinction.

Wolf

Wolverine *(Gulo gulo)* Arctic
Wolverines live in the cold northern forests of the taiga and tundra. They have an 80 centimetres long very strong body, long legs and large wide feet. They feed on mice and lemmings but prefer larger animals. The young are very playful.

Wolverine

Index

(Page numbers printed in italics refer to captions and illustrations)
See also page 78 for alphabetical illustrated list of species.